Reliability and Safety Assessment of Dynamic Process Systems

NATO ASI Series

Advanced Science Institutes Series

A series presenting the results of activities sponsored by the NATO Science Committee, which aims at the dissemination of advanced scientific and technological knowledge, with a view to strengthening links between scientific communities.

The Series is published by an international board of publishers in conjunction with the NATO Scientific Affairs Division

A	Life Sciences	Plenum Publishing Corporation
B	Physics	London and New York
C	Mathematical and Physical Sciences	Kluwer Academic Publishers Dordrecht, Boston and London
D	Behavioural and Social Sciences	
E	Applied Sciences	
F	Computer and Systems Sciences	Springer-Verlag Berlin Heidelberg New York
G	Ecological Sciences	London Paris Tokyo Hong Kong
H	Cell Biology	Barcelona Budapest
I	Global Environmental Change	

NATO-PCO DATABASE

The electronic index to the NATO ASI Series provides full bibliographical references (with keywords and/or abstracts) to more than 30000 contributions from international scientists published in all sections of the NATO ASI Series. Access to the NATO-PCO DATABASE compiled by the NATO Publication Coordination Office is possible in two ways:

- via online FILE 128 (NATO-PCO DATABASE) hosted by ESRIN, Via Galileo Galilei, I-00044 Frascati, Italy.

- via CD-ROM "NATO Science & Technology Disk" with user-friendly retrieval software in English, French and German (© WTV GmbH and DATAWARE Technologies Inc. 1992).

The CD-ROM can be ordered through any member of the Board of Publishers or through NATO-PCO, Overijse, Belgium.

Series F: Computer and Systems Sciences Vol. 120

Reliability
and Safety Assessment
of Dynamic Process Systems

Edited by

Tunc Aldemir

The Ohio State University, Nuclear Engineering Program
206 West 18th Avenue, Columbus, OH 43210-1107, USA

Nathan O. Siu

Center for Reliability and Risk Assessment, EG & G Idaho, Inc.
P. O. Box 1625, Mail Stop 2405, Idaho Falls, ID 83415, USA

Ali Mosleh

University of Maryland
Department of Materials and Nuclear Engineering
College Park, MD 20742-2115, USA

P. Carlo Cacciabue

Commission of the European Communities Joint Research Center
Institute for System Engineering and Informatics
I-21020 Ispra (VA), Italy

B. Gül Göktepe

Cekmece Nuclear Research and Training Center
P. K. 1 Havaalani, TR-34831 Istanbul, Turkey

Springer-Verlag
Berlin Heidelberg New York London Paris Tokyo
Hong Kong Barcelona Budapest
Published in cooperation with NATO Scientific Affairs Division

Proceedings of the NATO Advanced Research Workshop on Reliability and Safety Assessment of Dynamic Process Systems, held in Kusadasi-Aydin, Turkey, August 24–28, 1992

CR Subject Classification (1991): B.1.3, B.1.2, B.3.4, H.1.2, I.6.1, J.2, J.7

ISBN 978-3-642-08178-1

CIP data applied for.

© Springer-Verlag Berlin Heidelberg 2010
Printed in Germany

45/3140 - 5 4 3 2 1 0 - Printed on acid-free paper

Preface

Accumulating experimental, computational and historical evidence indicates that static approaches to system failure modeling, such as a conventional event-tree/fault-tree approach, may not be appropriate on a stand-alone basis for the reliability and safety analysis of dynamic process systems. Several methodologies have been proposed which explicitly account for the time element in system evolution to complement the event-tree/fault-tree approach. However, there is still substantial controversy over the need for such dynamic methodologies.

This book contains the proceedings of a NATO Advanced Research Workshop (ARW) on the Reliability and Safety Analysis of Dynamic Process Systems that was held in Kuşadası - Aydın, Turkey during August 24-28, 1992 and whose objectives were to:

- discuss the advantages and limitations of the methodologies proposed to date for the reliability and safety analysis of dynamic process systems,

- identify practical situations where dynamic methodologies could lead to significantly improved results, and,

- develop a benchmark exercise to compare dynamic methodologies with each other and with the conventional event-tree/fault-tree approach.

In order to achieve these objectives, both the developers of dynamic methodologies and also experts in the use of the conventional approaches for the reliability and safety assessment of dynamic process systems were invited to the ARW. The sessions were structured to encourage discussion of computational isues, as computational limitations are crucial in the application of dynamic methodologies to practical problems. Finally, since the operator is often an integral contributor to system dynamics, special emphasis was placed on human factors.

A total of 34 participants attended the ARW from 12 different countries. The participants represented 26 different institutions including universities, national laboratories, private consulting companies and regulatory bodies. Their combined expertise covered nuclear, chemical, mechanical, aerospace and defense systems. In view of the diverse methodological and systems background of the participants, the ARW began by reviewing the state-of-the-art through paper presentations in: a) dynamic methodologies, b) human reliability and dynamic systems analysis, and,

c) risk assessment for realistic decision making. These papers are provided in Parts 1 through 4 of this book. The review of dynamic methodologies discussed both issues and approaches and covered the DYLAM methodology, the theory of continuous event trees, several Markov model construction procedures, Monte Carlo simulation and utilization of logic flowgraphs in conjunction with Petri nets. The main focus of the presentations on human reliability and dynamic systems analysis was the modeling of the interaction between the dynamic plant environment and operator behavior. The influence of factors such as procedures and training or operational history on human behavior and the need to represent these influences through dynamic models were discussed. A number of operator modeling procedures, both currently employed and proposed, were also described. The presentations on risk assessment for realistic decision making concentrated on the modeling of dynamic effects with the conventional techniques and the need for improved methods in certain situations.

From the presentations and the discussions that followed, dynamic methodologies seem to be needed whenever there is complex hardware-human-process variable or hardware-software-process variable interaction in time. Examples include the response of a plant control room crew to an accident and the behavior of a software-based control system in maintaining a process variable at a setpoint. However, some participants felt that some of the situations described in the presentations could be handled through conventional techniques. While no strong consensus was reached during the ARW as to precisely when dynamic methodologies should be used, a fairly general consensus was reached that dynamic methodologies not only need to be further investigated but also need to be accelerated. In view of the makeup of the participants, this consensus was by no means a foregone conclusion (a number of the participants were experts in the application of conventional techniques to dynamic systems). The participants, as a whole, felt that the area was important and that the dynamic approaches being investigated were appropriate.

The need for and proposed format of the benchmark exercise were discussed in two half-day working sessions. The working session structures and a summary of the discussions is provided in Part 5 of this book. Although the benchmark problem was not fully defined, the basic characteristics of such a problem were identified, the strategy for the benchmark exercise was formulated and three committes were established to continue the formulation of a benchmark exercise program. It was agreed that only slight extensions of a simple problem proposed in the literature some years ago would provide sufficient detail to illustrate the usefulness of dynamic methodologies. It was also agreed that a phased approach should be employed to allow eventual incorporation of cognitive models for operators.

The NATO ARW summarized in this book represents a milestone in the discussion of the reliability and safety assessment of dynamic process systems. Besides providing a forum for the discussion of various approaches and techniques, an important outcome of this workshop was that it led to increased interest in dynamic methodologies on the part of the practitioners of conventional methodologies. Another important outcome was that it clarified misunderstandings regarding the need for and capabilities of various dynamic methodologies. A third important outcome was the previously mentioned development of an agreed upon strategy and action plan for developing a benchmark exercise program. Finally, due to the coincidence that a parallel NATO ARW on "Intelligent Systems: Reliability, Availability and Maintainability Issues" was also held in Kuşadası - Aydın, Turkey during August 24-27, 1992 (see NATO ASI Series F, Vol. 114, ed. O. Kaynak), a dialog was established with a portion of the reliability community interested primarily in approaches to the design of reliable systems, rather than the assessment of a system's reliability. The resulting interchange sensitized each group to the concerns of the other.

November 1993

The Organizing Committee

Tunc Aldemir (ARW Director)
The Ohio State University
Columbus, Ohio, USA

Nathan O. Siu
Idaho National Engineering Laboratory
Idaho Falls, Idaho, USA

Ali Mosleh
University of Maryland
College Park, Maryland, USA

P. Carlo Cacciabue
Commission of the European Communities Joint Research Center
Ispra (Varese), Italy

B. Gül Göktepe
Çekmece Nuclear Research and Training Center
Istanbul, Turkey

Acknowledgment

The organizers would like to thank NATO for providing most of the funds for the realization of the ARW on the Reliability and Safety Analysis of Dynamic Process Systems, as well these proceedings, and the institutions which indirectly contributed to the ARW by partially reimbursing the travel expenses of their representatives. The organizers would also like to acknowledge:

- Professor **Gürbüz Çelebi** of the Ege University, Bornova-Izmir, Turkey, for suggesting NATO as a potential funding source for such a meeting,

- Dr. **Ionannis Papazoglou** of the National Center for Scientific Research DEMOKRITOS, Athens, Greece, for his help in identifying potential participants from Europe,

- Professor **Okyay Kaynak** of the Boğaziçi University, Istanbul, Turkey and the director of the NATO ARW on "Intelligent Systems: Reliability, Availability and Maintainability Issues" (Kuşadası - Aydın, Turkey, August 24-27, 1992) for his cooperation in making both ARWs more interesting and fruitful, and last, but definitely not the least,

- Ms. **Demet Sakarya**, the ARW secretary, for her excellent support during the meeting.

Contents

Part 1

Dynamic Approaches - Issues and Methods

Dynamic Approaches - Issues and Methods: An Overview

Nathan O. Siu

Center for Reliability and Risk Assessment, Idaho National Engineering Laboratory
Idaho Falls, ID 83415, USA

Abstract. Conventional risk and reliability assessment methodologies are static in nature and have potential weaknesses when applied to dynamic process systems. An overview of these weaknesses, issues that need to be dealt with in a dynamic analysis and methodologies that have been proposed to address these issues is provided.

Keywords. Risk assessment, event tree, fault tree, dependent failures, dynamic methodologies

1 Background

An important characteristic of many engineering systems is that they behave dynamically, i.e., their response to an initial perturbation evolves over time as system components interact with each other and with the environment. Conventional analyses of system risk and reliability, on the other hand, frequently do not address this characteristic. The well-known event tree/fault tree approach [1] typically used in these analyses treats accident scenarios as static sets of safety barrier (top event) successes and failures; it does not literally simulate system response to an initiating event. (Note that event tree/fault tree analyses are often supported by limited, offline dynamic analyses.) The effects of process variable behavior and operator actions on scenario development are incorporated through the success criteria defined for the event tree/fault tree top events.

The event tree/fault tree methodology is well-suited in many situations for identifying logically correct, or, at least, conservative relationships between top event successes and failures and plant damage states. However, as argued more extensively in [2], the methodology has some potential weaknesses when quantifying the risk associated with scenarios for which the plant dynamic behavior is a significant factor.

To explore this issue, recall that the key issue in scenario quantification is the assessment of the probability of multiple failures, since current "defense-in-depth" designs are typically resistant to single failures. This means that great care is required in identifying and quantifying dependencies between the failure events. The event tree/fault tree methodology directly addresseses multiple failure dependencies associated with: 1) common cause initiating events (e.g., earthquakes), 2) functional dependencies (e.g., support system failures), and, 3) shared equipment dependencies (e.g., when a single basic event appears in multiple top events). The methodology, does not, however, provide strong support in identifying and analyzing dependencies that fall outside of these three categories. These other dependencies involve situations where the status of the plant cannot be defined solely in terms of top event successes and failures. Of particular interest to these proceedings are dynamic scenarios whose development is strongly affected by automatic control systems or operator actions.

The problem is that the event tree/fault tree methodology represents each accident scenario as a set of hardware failures and operator errors. The latter are treated in much the same fashion as hardware failures, and often treated at a very broad level, e.g., failure to depressurize the reactor coolant system in τ minutes. As a result, many of the conditions affecting control system actions and operator behavior (e.g., behavior of plant process variables, previous decisions by the operating crew) are not explicitly included in the model. This affects the assessed level of dependence between events. For example, probabilistic risk assessment (PRA) models rarely identify risk significant situations in which operators turn off needed safety systems, although this was a prime contributor to the TMI-2 accident. In the absence of a context provided by a description of the dynamic progression of the accident, it is difficult for an analyst to develop appropriate conditional frequencies for such events.

As a related point, the treatment of human error in a fashion analogous to the treatment of hardware failure inhibits accurate modeling of the remainder of the accident sequence following an error. The likelihood that an operating crew fails to perform a required task correctly (within a given amount of time) is treated explicitly. However, the different ways in which the crew may perform the task incorrectly, and the resulting dynamic responses of the plant/crew system to these different errors, are not treated. Therefore, the proper boundary conditions for establishing the conditional failure probabilities for top events downstream of the task performance failure are not provided. Moreover, from the standpoint of risk management, the lack of realistic treatment of the scenario following human error can lead to an incomplete identification of factors important to risk, and of alternatives that can be employed to reduce risk. A similar argument applies to the treatment of control (software as well as hardware) systems using binary success/failure models.

In [2] it is pointed out that in order to address these weaknesses, it must be recognized that: a) control systems (including operators) and plant components are interacting parts of an overall system that responds dynamically to upset conditions, b) the actions of control systems (operators) are dependent on their knowledge concerning the current state of the plant, and c) operators have memory; their beliefs at any given point in time are influenced (to some degree) by the past sequence of events and by their earlier trains of thought. If these observations are important for the scenario being analyzed, a modeling structure significantly from that of current event trees and fault trees is needed. Such a model must carry information on the following:

- Current hardware status.
- Current levels of process variables.
- Current control system/operator knowledge concerning plant state.
- Scenario history.
- Time.

The event tree/fault tree methodology explicitly treats the first item and can deal with the fourth item to some extent. The remaining items affect the behavior of automatic control systems, the operators, or both. (Time influences the process variable calculations, and can also be a key performance shaping factor when modeling operator behavior.)

2 Methodologies

Three dynamic analysis modeling approaches developed in recent years to address the previously discussed problems with conventional event tree/fault tree analyses are:

- Dynamic event tree methodology
- Discrete state-transition modeling
- Event simulation (analog Monte Carlo)

The dynamic event tree methodology, as implied by its name, extends the notion of an event tree to treat event sequence branching over time. This approach requires the explicit tracking of possible scenarios. Each scenario track provides the contextual information needed to determine the likelihood of system state changes during any point in the scenario. (The system state is generally defined implicitly in terms of the states of the system "components" - including operators.) The earliest and best-known application of the dynamic event tree methodology is the dynamic logical analytical methodology (DYLAM), whose latest version (DYLAM-3) is described in [3]. DYLAM-3 treats both deterministic and stochastic component state transitions in a general manner. In [3] it is shown how this approach can be used to treat a simple holdup tank problem introduced in [4], and how the results of this analysis compare with those obtained using conventional fault tree methods. DYLAM-3 can be used to model operator behavior in a limited fashion; an extension of DYLAM designed to deal explicitly with dynamic crew/plant interactions is described in [5].

In contrast with the dynamic event tree approach, discrete state-transition modeling requires the explicit *a priori* identification of discrete system states. (In cases where process variables must be tracked, this approach requires that continuous process variable ranges be discretized as well.) The dynamic development of an accident is then represented by deterministic or probabilistic transitions between states. A discrete state-transition model can represent an accident in a much more compact form (a state-transition matrix) than a dynamic event tree; however, history-dependent information (i.e., how the system happened to arrive at a particular state at a particular point in time) can be more easily lost. For this reason, the state-transition approach is typically employed in Markovian analyses, where knowledge of the current system state is assumed to be sufficient for determining the likelihood of future state transitions. In [4] an early analysis is described of a simple controlled system combining discrete (hardware-component related) and continuous (process-variable related) variables in a Markovian state-transition model. In these proceedings, [6] describes the advantages and disadvantages of Markovian state-transition modeling for reliability analysis in comparison with conventional methodologies, while [7] describes a qualitative model useful for addressing the time-dependent behavior of embedded software systems.

The dynamic event tree and the discrete state-transition approaches are clearly related in that both require the analyst to identify discrete component/system states and possible transitions between states. This must be performed even for situations for which the definition of such states may be cumbersome (e.g., when treating operator behavior). Event simulation (also called analog Monte Carlo) is a Monte Carlo-based modeling approach that can relieve some of this burden.

The event simulation approach, discussed more thoroughly in [2] and [8], requires the analyst to develop a model whose elements and events correspond directly with those of the system. Because of its analogous relationship with the system, the analyst can better treat extremely complex interactions among objects; the resulting model is also more easily interpretable. The primary difficulty with the event simulation approach is numerical; intelligent sampling schemes must be devised to accurately deal with the rare events of interest to risk and reliability analysis. (Of course, the dynamic event tree and state-transition methodologies also have numerical difficulties, due to the basic complexity of the problem addressed. These difficulties are simply manifested in different ways: multiple paths in a dynamic event tree or a large transition matrix in a state-transition model.) In [8] the event simulation approach is discussed in connection with the holdup tank problem introduced in [4]. It provides useful discussion on when dynamic methods are likely to lead to different results as compared with conventional approaches, and also discusses the need for caution in employing a commonly used sampling scheme, a caution that arises precisely because the sampling scheme does not account for the dynamic behavior of the system.

3 Closing Remarks

The methodology categorization used in the preceding discussion is useful for distinguishing the approaches of different investigators over the years, but does not strongly emphasize differences in the basic characteristics of the methodologies. Reference [9] presents a systematic categorization scheme based on a consideration of modeling issues (e.g., process variable discretization versus time discretization) and numerical algorithms (e.g., explicit transition matrix multiplication versus Monte Carlo simulation).

Reference [9] also presents a general approach for analyzing the behavior of probabilistic dynamic systems and shows how the dynamic event tree, state-transition modeling, and event simulation approaches discussed above represent special instances of the general approach. Strong arguments are made concerning the theoretical advantages of Monte Carlo simulation over other methods when dealing with complex situations. Additional work is needed to determine if these advantages are realized in practical analyses (where model development resources can be considerable).

References

1. U.S. Nuclear Regulatory Commission: Reactor safety study. NUREG-75/014/WASH-1400 (1975).

2. Siu, N.: Risk assessment for dynamic systems: an overview. To appear in Rel. Eng. and Sys. Safety, accepted for publication (1993)

3   Cojazzi, G., Cacciabue, P. C.: The DYLAM approach for the reliability analysis of systems with dynamic interactions. These proceedings

4   Aldemir, T.: Computer-assisted markov failure modeling of process control systems. IEEE Trans. Rel. R-36, 133-144 (1987)

5. Acosta, C., Siu, N.: Dynamic event trees in accident sequence analysis: application to steam generator tube rupture. Rel. Eng. and Sys. Safety, accepted for publication, 1993

6. Papazoglou, I. A.: Markov reliability analysis of dynamic systems. These proceedings

7. Apostolakis, G.: Dynamic safety analysis of embedded systems. These proceedings

8   Marseguerra, M., Zio, E.: Approaching dynamic reliability by Monte Carlo simulation. These proceedings

9. Devooght, J., Smidts, C.: Probabilistic dynamics: the mathematical and computing problems ahead. These proceedings

The DYLAM Approach for the Reliability Analysis of Dynamic Systems

Giacomo Cojazzi, Pietro Carlo Cacciabue

Commission of the European Communities, JRC, Institute for Systems Engineering and Informatics, 21020, Ispra, Varese, Italy

Abstract. The evolution of incidental sequences in a system is the combined result of stochastic and deterministic events, the former due to the failures of the components, the latter related to the physical behaviour of the system. The DYLAM-3 code for the reliability assessment of a system represents a powerful tool for integrating deterministic and failure events and it is essentially based on the systematic simulation of the physical process under study, both in nominal and in failed conditions. This paper describes the main features of the DYLAM-3 code with reference to a simple system already analyzed by other dynamic approaches. The same system is also studied with a time dependent fault tree approach in order to show some features of dynamic methods.

Keywords. Reliability analysis, dynamic systems, physical modelling, fault tree/event tree methods, probabilistic techniques.

1 Introduction

Classical reliability techniques separate the probabilistic analysis from the study of the consequences, so that it becomes very difficult to assess the time dependent probability of a system top event, when complex dependencies exist, and when the physical evolution of the system cannot be de-coupled from its probabilistic behaviour. Moreover, proper modeling of the time dimension adds additional complexity to the problem: it seems more correct to speak, in these cases, of dynamic interactions instead of dependencies. Typical problems are those due to the presence of control loops, human actions and/or interventions of the protection systems (Cojazzi et al. 1992a). Moreover, actual situations often present a combination of such problems. In the past 10 years, different methodologies have been proposed in order to improve the conventional Event Tree-Fault Tree (ET-FT) approach. These methodologies make use of the knowledge deriving from more or less detailed time dependent plant models, in order to better describe the probabilistic evolution of the system in time.

Siu (1992) gives a rather comprehensive description of the limits of the ET-FT approach. He also presents an overview of the existing methods for improving the Probabilistic Safety Assessment (PSA), starting from simple extensions of the ET-FT framework, to arrive to truly dynamic methods which employ a model of the physical system. It is beyond the scope of this paper to present a complete list of the different approaches that have been proposed. We will only recall here the most important endeavors in this domain, which has been initially tackled by Amendola and Reina in 1981, with the first version of the DYLAM methodology. Aldemir (1987, 1989), Aldemir et al. (1992), and Limnios and Cocozza (1992) model the stochastic-deterministic nature of dynamic systems as a Markov chain, after a suitable discretization of the process variables. Deoss and Siu (1989) have proposed a simulation oriented approach employing Monte Carlo techniques for introducing failures in time. Acosta and Siu (1991) have also developed a Dynamic Event Tree Analysis Methods (DETAM) which maintains the event tree structure and is similar to the DYLAM methodology. Devooght and Smidts published (1992a) a general theory for describing the deterministic and stochastic nature of incidental events and employed Monte Carlo techniques in order to study a fast reactor transient (1992b). Monte Carlo techniques are also being applied by Marseguerra and Zio (1992).

In this paper, we will briefly recall the background underlying the development of the DYLAM approach and how this has been implemented in a general purpose code that is now available. A sample case from the literature (Aldemir 1987), consisting of an holdup tank and the related supply system, is analyzed with DYLAM. The same system, with minor modifications is extensively analyzed with the classic FT methodology and with DYLAM in order to present some features of the DYLAM approach for the reliability analysis of systems.

2 The DYLAM Methodology

The basic idea of the DYLAM methodology (Amendola and Reina 1981) is to provide a tool for coupling the probabilistic and physical behaviour of a system for more detailed reliability analysis. All the knowledge of the physical system under study is contained in the numerical simulation where the components of the system are modeled in terms of different working states (nominal, failed on, failed off, stuck, etc.). Once the simulation program is linked to the DYLAM code, this takes into account the time history of the logical states of the components, by assigning initial states and triggering random transitions in the component states. One characteristic of DYLAM is to follow all the different paths resulting from the transitions in time of the component states and to drive the corresponding simulations. For each path a time dependent probability is evaluated, so that the probability of occurrence of certain top event is obtained by adding the probability of the corresponding sequences. Because of its dynamic features, the DYLAM analysis can be seen as complementary to the ET-FT techniques when the detailed modeling of complex scenarios or the assessment of time dependent top probabilities is needed.

Different applications of the DYLAM code have already been published in literature (Cacciabue et al. 1986, 1991, 1992a), including the analysis of subsystems of a real nuclear power plant. Very recently a new release of the code (DYLAM-3) has been issued which contains a more precise probabilistic treatment (Cojazzi et al. 1992b).

DYLAM offers a framework to take into account different kind of probabilistic behaviours. In particular, for each component of the system, the user can choose between six different probabilistic options, namely:

0. Constant probabilities for initial events and component states.
1. Stochastic transitions between the states of the component.

2. Functional dependent transitions, for failures on demand and physical dependencies.
3. Stochastic and functional dependent transitions.
4. Conditional probabilities for dependencies between states of different components.
5. Stochastic transitions with variable transition rates, i.e. function of time or process variables.

With DYLAM it is possible to trigger initiating events involving arbitrary combinations of failures or to perform exhaustive analysis involving the systematic combinations of initial failures up to a prefixed order. Concerning its dynamic capabilities, DYLAM can easily treat transitions (failures and repairs) in-time with constant (option 1) or variable transition rates (option 5), as well as failure on demand (option 2) and failures on demand combined with transitions in time (option 3).

In order to briefly present the capabilities and drawbacks of the method, we list hereafter all the main assumptions underlying the DYLAM code, the mechanisms for handling the branchings between different sequences and the stopping and limiting rules.

- DYLAM is not restricted to binary logic.
- The time step of analysis Δt, is fixed and thus branchings can happen only at discrete times.
- The time step Δt must be adequate to the time constants of the physical system. Moreover Δt must be less than the minimum mean life time of each state of the system.
- All sequences generated by DYLAM have the same duration (Tmax); transitions in time should be considered for a component or system only if its mean life time in nominal state is comparable to the mission time.
- Branching in time may be due to components described by options 1, 2, 3 and 5. In particular for options 1, 3, and 5 a probabilistic threshold applies. Whenever the probability of the sequence drops below a predefined percentage of the initial value (Wlim) following a failure of a component, a branching applies to that component. Options 2 and 3 cause a branching whenever the physical or logical variable associated to the component changes of interval. Notice that option 3 has both the mechanisms. More complex dependences can be described by suitable modeling in terms of DYLAM components.
- Simultaneous multiple failures are never taken into account.
- Sequences with a probability falling under a predefined threshold (Plim) are discarded.
- Absorbing conditions can be assigned to the system as a whole: when the system reaches a particular state (safe or unsafe), further branchings are not considered for that sequence.
- Absorbing states can be introduced for characterizing the states of each component.

3 The Reservoir Problem

In order to better understand the features of the DYLAM code when applied to the reliability analysis of a dynamic system, we will consider a simple case study adapted from Aldemir (1987). The system consists of 1) a storage tank, whose level has to be kept in a suitable interval, 2) one supply system, with a main inlet pump (PUMP1) and an additional safety pump (PUMP2) and 3) a release valve located on the bottom of the tank (Fig. 1).

At time t=t0 the initial level of the tank is zero and in nominal conditions the amount of liquid entering the tank by PUMP1 equals the amount of liquid released by the valve. The nominal level is therefore maintained constant until failures in time happen to at least one of the three components: if one component fails the level of the tank varies in time and eventually the control system intervenes in order to keep the level constant. The aim of the analysis is to evaluate the time dependent probability of Overflow (level > 3m) and of Dryout (level < -3m). Hereafter we list all the assumptions and the data common to all the numerical results hereafter described:

1. The liquid level is initially zero. The components and their control systems are independent.
2. All the components are in nominal conditions at time zero. The mean failure time of the components are $\tau_1 = 219$ h, $\tau_2 = 175$ h and $\tau_3 = 320$ respectively for PUMP1, PUMP2 and the VALVE. All failed states are considered absorbing.
3. In normal condition the mass flow rates of PUMP1 and of the VALVE are the same, so that, if nothing happens, the level remains constant at level zero, the initial one.
4. The mass flow rate of PUMP2 is half of PUMP1 and of the VALVE (as in case F of Aldemir, 1987).
5. The cross section of the tank is assumed unitary and constant and the density of the liquid is also assumed constant, therefore the mass flows can be expressed in meters per hours: PUMP1 and VALVE give a rate of level change of 0.60 meters per hours.

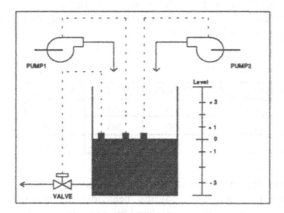

Fig. 1. The tank scheme (From Aldemir 1987, and Siu 1992)

4 Numerical Results

Two different situations have been analyzed, the differences being in the control laws for the components and in the definition of the failed states. These two situations are referred to as case 1 and case 2 and are described in detail in the next sections.

4.1 Results for Case 1

The first case considered is the same as the one presented by Aldemir in 1987. Table 1 recalls the control laws adopted, where each component has just two states, the nominal and the failed; in the nominal case each component is "on" or "off" according to the respective control laws. In this case it is assumed that the failures of a component are actually failures of the corresponding control unit: once failed each component behaves opposite to the control laws, so that the failure modes depend on the physical system behaviour, and an estimate of the overflow and dryout probabilities without considering the dynamics of the system is almost impossible. As an example, let us consider that the component PUMP1 fails in the control region 2, i.e. between -1 and +1 m: since in this region it should be "on" this means that it will be failed off. Briefly, the method proposed by Aldemir employs a differential equation for describing the level of the tank and failure data in order to construct a unique Markov chain for describing the probabilistic behaviour of the system in discrete time and in the discretized controlled variable state space.

The same system has been also studied by Deoss and Siu (1989) with their method DYMCAM (DYnamic Monte Carlo Availability Model) which is based on the simulation of the system and employs Monte Carlo techniques for introducing failures in time. The control laws employed are the same used by Aldemir, the only difference is that a more component oriented approach is considered with two failed states for each component: a failed on (open) and a failed off (closed). Deoss and Siu developed also a simplified Markov model of the system with the additional assumption that if a component fails in a control region a second failure does not occur untill the system has entered a new control region. In figure 2.a the time dependent overflow and dryout cumulative probabilities are reported for a) the simplified Markov method (continuous line), b) Aldemir (symbols) and c) Deoss and Siu methods (dashed line). The difference of the results of Deoss and Siu and Aldemir is mainly due to the different definitions considered for the failed states.

Table 1 Control laws of the system in case 1

Region	Liquid Level Z	PUMP1	PUMP2	VALVE
1	$L < -1$	on	on	close
2	$-1 \leq L < 1$	on	off	open
3	$1 \leq L$	off	off	open

The same system was also analyzed with the DYLAM code and a very simple routine was written in order to simulate the physical behaviour the system (see the next section). The failure modes are those considered by Aldemir, i.e. when a component fails it behaves opposite to its control laws.

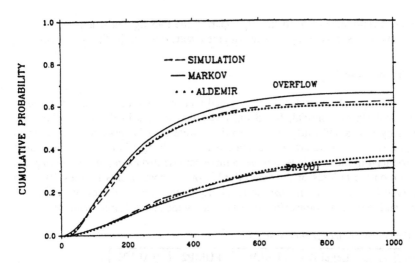

Fig. 2.a. Overflow and Dryout probabilities with simplified Markov (continuous), Aldemir (symbols) and Siu (dotted), (from Deoss, Siu,1989).

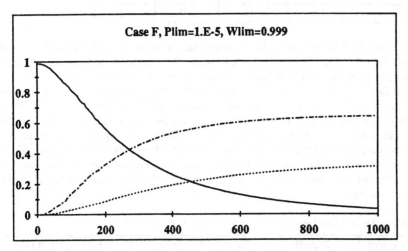

Fig. 2.b. DYLAM estimates: overflow (dot-dash), dryout (dotted) and nominal probabilities (continuous).

The time dependent Overflow and Dryout probabilities were evaluated during a period of 1000 hours and with a time step of analysis of one hour. The cut-off probability (Plim) is 1.e-05 and the threshold level is 0.999 (Wlim): this means that, whenever the probability of the sequence becomes less or equal a fraction Wlim of its initial value, a branch point is considered. Moreover when the probability of a sequence becomes less or equal to Plim, the sequence is discarded. In average one out 100 of these discarded sequences is maintained with a probability multiplied by 100 in order not to lose probability. The results are reported in figure 2.b and were obtained on a Sun 4/75 workstation, the CPU time was approximately of 5500 s. The continuous line represents the success probability, the overflow is the dashed line and the dryout probability is reported in dotted line. Comparing the two figures it appears that the DYLAM estimate is in good agreement with the results reported by Deoss and Siu.

4.2 Results for Case 2.

In the second example considered we assumed that the three components may be in three different states, nominal, failed on, and failed off (as in the tank study case performed by Deoss and Siu). The control laws were slightly modified according to the rules of Table 2. There are several reasons for this change: 1) the two failed states were introduced in order to have a more component oriented example, in this case the focus is not on the failures of the controller but on the components failures; 2) by so doing it is also possible to analyze the reliability of the tank by the conventional FT technique and to present some differences between static and dynamic methods.

Table 2. Control laws of the system in case 2

Region	Liquid Level Z	PUMP1	PUMP2	VALVE
1	$L < -1$	on	on	close
2	$-1 \leq L < 1$	-	-	-
3	$1 \leq L$	off	off	open

As far as the different control laws, notice that Table 2 can be obtained from Table 1 simply by removing the rules relating to region 2. In order to understand the effect of this, imagine, for example, that a failure in one component happens, e.g. the VALVE failed open: the level of the tank will decrease down to level -1 before the control system activates the auxiliary safety pump (PUMP2). With the laws of Table 2, once the second pump is activated, it will remain switched on up to the level of the liquid will reach +1 m, then it will be switched off. By so doing all the intermediate region becomes, from the control point of view, a dead band in which the level of the liquid is allowed to fluctuate in between. Table 3 summarizes all the nominal and failed states of the components, together with the reliability data employed in the DYLAM analysis.

Table 3. Nominal and failed states of the components and transition rates

Component	State 0	State 1	State 2	Transitions and rates
PUMP1=U^1	nominal $K_1=1$ if on $K_1=0$ if off	failed-on $K_1=1$	failed-off $K_1=0$	$\lambda^1_{01}=\lambda^1_{02}=1/(2*\tau_1)$ else $\lambda^1_{ij}=0$
PUMP2=U^2	nominal $K_2=1$ if on $K_2=0$ if off	failed-on $K_2=1$	failed-off $K_2=0$	$\lambda^2_{01}=\lambda^2_{02}=1/(2*\tau_2)$ else $\lambda^2_{ij}=0$
VALVE = U^3	nominal $K_3=1$ if on $K_3=0$ if off	failed-open $K_3=1$	failed-closed $K_3=1$	$\lambda^3_{01}=\lambda^3_{02}=1/(2*\tau_3)$ else $\lambda^3_{ij}=0$

As usual in the DYLAM methodology, when a component is in the nominal state its actual condition depends on control laws, this means that a nominal component may be active or switched off as consequence of the level of the tank. Notice that all failed states are assumed absorbing so that repair rates are not considered. It can be easily seen that, in the case of only one failure of any component, the system is completely controllable so that none of the top conditions (overflow or dryout) can arise from a single failure.

A simple routine linked with the DYLAM code evaluates the level of the tank as a consequence of failures of the components and of the intervention of the control system:

$$\begin{cases} L(t+\Delta t)=L(t)+Q_{net}\cdot\Delta t \\ Q_{net}=K_1\cdot Q-K_2\cdot Q+K_3\cdot\dfrac{Q}{2} \end{cases} \tag{1}$$

Here Q is the rate of level change due to the PUMP1 and the VALVE and is equal to 0.6 m/h and the K_i coefficients ($i=1,2,3$) are automatically assigned by DYLAM according to Table 3. Moreover $L(t=0)=0$.

This case can be analyzed with a fault tree technique in order to find the minimal cut sets of the top conditions overflow and dryout. It can be easily found that, for the overflow there are three minimal cut sets of order two (C_i, $i=1,2,3$) and that for the dryout there is only one minimal cut set of order two (C_4). Therefore:

$$P_{ov}=P(C_1\cup C_2\cup C_3)=$$
$$=\sum_{i=1}^{3}P(C_i)-P(C_1\cap C_2)-P(C_1\cap C_3)-P(C_2\cap C_3)+P(C_1\cap C_2\cap C_3) \tag{2}$$

$$P_{dr}=P(C_4) \tag{3}$$

where

$$C_1=U_1^1\cap U_1^2;\quad C_2=U_1^2\cap U_2^3;\quad C_3=U_1^1\cap U_2^3;\quad C_4=U_2^1\cap U_1^3 \tag{4}$$

and U_j^i, $i=1,...3$; $j=0,...2$ represents the event "component i in state j".
Inserting eqs. (4) in eq. (2) and simplifying, yields:

$$P_{ov}=P(C_1\cup C_2\cup C_3)=\sum_{i=1}^{3}P(C_i)-2P(U_1^1\cap U_1^2\cap U_2^3) \tag{5}$$

Expressions (3) and (5) can be employed in order to evaluate the probability of overflow and dryout of the tank in a FT scheme once the elementary probabilities $P(U_j^i)$ are known. Since no repairs are considered:

$$P[U_0^i(t)]=e^{-(\lambda'_{01}+\lambda'_{02})t} \qquad\qquad i=1,2,3 \tag{6}$$

$$P[U_j^i(t)] = \frac{1}{2}\left[1 - e^{-(\lambda'_{01}+\lambda'_{02})t}\right] \qquad\qquad j=1,2 \qquad\qquad (7)$$

These expressions will be applied in the comparison between the FT and the DYLAM results.

4.2.1 Failures Concentrated at the Initial Time of the Analysis

First of all, we can consider a simple comparison between the results obtained with the FT technique and with DYLAM in the case that failures do not happen in time but are introduced only at time zero. To this aim it is necessary to fix the duration of the time interval to be considered in order to convert the failure rates of the components in suitable probabilities of failure. A time interval $\Delta t=24$ hours is considered and all the possible failures are lumped at time zero with the probability they reach at time $t=24$:

$$P[U_0^i(t=0)] = e^{-(\lambda'_{01}+\lambda'_{02})\Delta t} \qquad\qquad P[U_{j\neq0}^i(t=0)] = \frac{1}{2}\left[1 - e^{-(\lambda'_{01}+\lambda'_{02})\Delta t}\right] \qquad (8)$$

The elementary probability data thus obtained are employed both in the DYLAM analysis, as input data, and in the FT analysis (eqs. 2 and 5) and the results are reported in figure 3.a. The results of the FT analysis are presented with symbols and the results of DYLAM are presented with lines. In particular the dotted line is the DYLAM estimate of the dryout probability, the overflow probability is reported in dot-dash line and the FT estimate for the dryout probability is displayed by the triangles. For the overflow probability with the FT we present two results, the exact one represented by the diamonds, which correspond to eq. (5) and an approximate one employing the rare event approximation, usually adopted in FT analysis, which corresponds to the first term at the right hand side of eq. (5) and is displayed with the squares.

By comparing the FT results and the results of DYLAM it appears that, for the DYLAM estimate, the probability of the overflow event is greater then zero only for $t > 4$ h and the dynamics of the dryout event is even slower. Obviously once the situation is in steady state the FT and DYLAM results are coincident. The FT approach, neglecting the dynamics of the system assumes simultaneity between the occurrence of the cut sets and the occurrence of the top event, while a simulation oriented approach like DYLAM evaluates correctly the delays due to the time constant of the physical system and produces less conservative results.

The DYLAM results have been obtained by the use of the *complete mode*, in which DYLAM generates and then analyzes the incidental sequences resulting from all the possible combinations of initial failures in the system.

The total number of incidental sequences considered by DYLAM is therefore: $N^1 \times N^2 \times N^3 = 27$, where N^i represents the number of states of the i-th component. DYLAM consider first the *nominal sequence*, i.e. the one in which all the components are in nominal conditions ($U_0^1 \cap U_0^2 \cap U_0^3$), then it considers the *sequences of order 1* (6 sequences), i.e. the ones in which only one component is failed, e.g. PUMP1 is failed-on ($U_1^1 \cap U_0^2 \cap U_0^3$), then the sequences of order two (12 sequences) are analyzed and finally the sequences of order three are analyzed, in which all the three components are failed (8 sequences). For each of these incidental sequences the level of the liquid in the tank is evaluated, according to eq. (1) and at each time step of analysis the code checks if the overflow or tryout levels are reached. DYLAM adds automatically the probability of the *top sequences* to the overall top probabilities. Notice that this is possible since the DYLAM sequences are mutually exclusive by construction. In this case none of the sequences is dropped since no cut off rules are applied. In particular there are 7 different sequences resulting in the overflow condition and 3 sequences that give raise to a dryout event. Notice that, due to the dynamics of the system these sequences result in the respective top conditions at different time instants: this is the reason of the step-wise behaviour of the figure 3.a. Hereafter we report the steady state overflow and dryout probabilities as estimated by DYLAM. Notice that they are equivalent to eq.s (5) and (3) respectively:

$$P_{ov} = P(U_1^1 \cap U_1^2 \cap U_0^3) + P(U_1^1 \cap U_0^2 \cap U_2^3) + P(U_0^1 \cap U_1^2 \cap U_2^3) + P(U_1^1 \cap U_1^2 \cap U_1^3) +$$
$$P(U_1^1 \cap U_1^2 \cap U_2^3) + P(U_1^1 \cap U_2^2 \cap U_2^3) + P(U_2^1 \cap U_1^2 \cap U_0^3) \tag{9}$$
$$P_{dr} = P(U_2^1 \cap U_0^2 \cap U_1^3) + P(U_2^1 \cap U_1^2 \cap U_1^3) + P(U_2^1 \cap U_2^2 \cap U_1^3)$$

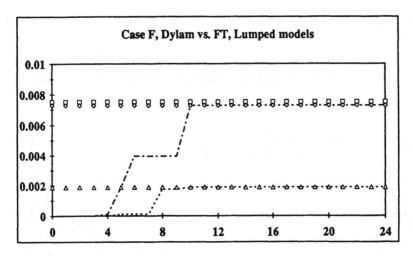

Fig. 3.a. Overflow and Dryout probabilities with FT (symbols) and with DYLAM (lines) in the case of failures only at time t=0.

4.2.2 Stochastic Failures

A more precise evaluation of the overflow and dryout probabilities can be obtained by considering the failures of the components as stochastic events exponentially distributed during the time of interest. Figure 3.b compares the results that can be obtained with a static and a time dependent FT analysis. In particular the static results correspond to the lumped model of failure and are shown again in figure 3.a, while the time dependent results are obtained with eq.s (2) to (6) for increasing values of time, with a time step of one hour. Again diamonds represent the cumulative distribution function (cdf) of overflow vs. time, and triangles are associated to the dryout cdf.

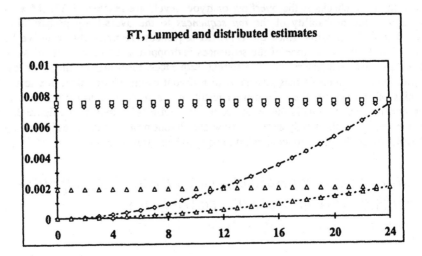

Fig. 3.b. Overflow and Dryout probabilities with FT; cases of failures only at t=0 (symbols) and of failures in time (line+symbols).

The time dependent overflow and dryout cumulative probabilities obtained with the FT are compared with the corresponding time dependent DYLAM estimates in figures 4.a and 4.b.

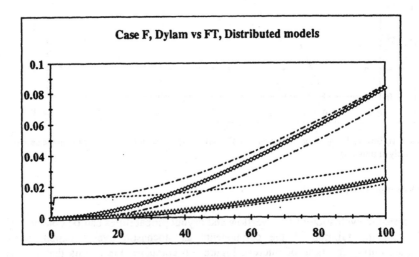

Fig. 4.a. Overflow and Dryout probabilities with FT (symbols) and upper and lower bounds estimates with DYLAM (lines) in the case of stochastic failures, Δt DYLAM is 1 h.

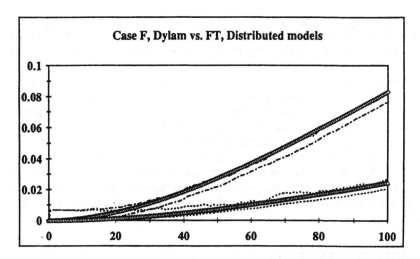

Fig. 4.b. Overflow and Dryout probabilities with FT (symbols) and upper and lower bounds estimates with DYLAM (lines) in the case of stochastic failures, Δt DYLAM is 0.5 h.

In particular, in figure 4.a the FT results are reported with symbols (diamonds for overflow, triangles for dryout) and the DYLAM results are represented (the overflow) by the lowest dot-dashed line and (the dryout) by the dotted line (the lowest of the two lines).

In the DYLAM analysis it is assumed the all the three components are initially in nominal conditions and that they may undergo stochastic failures according to the failure rates (Table 3). The DYLAM time step of analysis is of one hour and the branching parameter (Wlim) is set equal to 0.999, which means that, whenever the probability of the system to remain in the previous state becomes less then a fraction Wlim of the initial value, a failure is inserted in one component. In practice this value of Wlim is such that, at each time step of analysis, each of the three components may fail. This is necessary in order to get correct results. By so doing the number of event sequences becomes unmanageable and, for practical reasons, a cut off threshold probability (Plim) equal to 10^{-6} is set. However, as it has been done for case 1, in order to ensure completeness of the events, 1 % of these low probability sequences have been selected on a random basis, with increased probability by a factor 100. The other lines represent, respectively, the upper bounds for the DYLAM estimate of the overflow probability (the dot-dash) and of the dryout one (the dotted line).

In the DYLAM analysis we considered all the three mutually exclusive events of overflow, dryout and success, at any time the sum of the probabilities of these three exhaustive and mutually exclusive events should be unity. This normalization to unity in not always achieved, in particular the algorithms embedded in DYLAM do not consider multiple simultaneous transitions. Due to this effect, the probability corresponding to multiple transitions is assumed lost and constitutes an uncertainty value that could be related to any of the top events analyzed. In other words, if $P_{ov}(t)$ is the DYLAM estimate of the overflow probability, corresponding to the lower dot-dashed line and $P_{lost}(t)$ is the probability corresponding to the sequences not considered by DYLAM, the upper bound for the DYLAM estimate of the overflow is computed as

$P_{ovub}(t)=P_{ov}(t)+P_{lost}(t)$. The same considerations apply to the dryout probability. Therefore the couples of the dot-dash and dotted line represent the uncertainty affecting the DYLAM results, and it is clear that in this case there is not a significant difference between the FT and the DYLAM results

Since the probability loss, and therefore the uncertainty in the results, is due to the neglection of multiple simultaneous transitions, it is clear that, at the decreasing of the time step of analysis, the uncertainty affecting the results will be smaller. This can be seen from figure 4.b which reports the results of DYLAM obtained with a time step of half an hour. It appears that the uncertainty affecting the results is smaller. However, also in this case there is no significant difference between the results of the two approaches. The reason is that in this simple case the only element active from the dynamic point of view is due to the presence of the tank, which acts as delay element, and since this effect is small (due to reduced capacity of the tank), it can not be fully appreciated.

4.2.3 Effect of the Tank Capacity

In order to present the effects due to the variation of the tank capacity on the overflow and dryout probabilities, we considered two different systems consisting of the same tank already analyzed but with two different increased capacities. In one case the overflow level is increased up to 12 meters (C=12) and in the second case the overflow level is set at 20 meters (C=20). It is intuitive that the cumulative probability of overflow will be progressively smaller as the capacity of the tank increases.

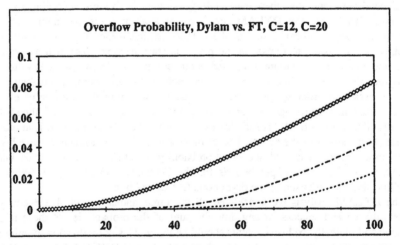

Fig. 5.a. Overflow probability with FT (symbols) and with DYLAM in the case of stochastic failures and overflow level of 12 meters (dot-dash) and 20 meters (dotted).

The results obtained with DYLAM are reported in figures 5.a and 5.b respectively for the overflow and dryout events, together with the corresponding estimates obtained with the FT approach which is clearly insensitive to the variation of the tank capacity. For simplicity the DYLAM results are reported without the respective error band. In figure 5.a the diamonds represent the FT estimate, the dot-dashed line corresponds to the tank with 12 meters of overflow level, and the dotted line refers to the overflow level of 20 meters. Notice that in this case the DYLAM results are significantly different from the FT ones. From figure 5.b it appears that the dryout probability is not affected by the variation of the overflow level of the tank. This is because no repair events are considered and thus no sequences resulting in the overflow may turn later in dryouts and vice-versa. Therefore the variation of the overflow level does not affect the dryout probability.

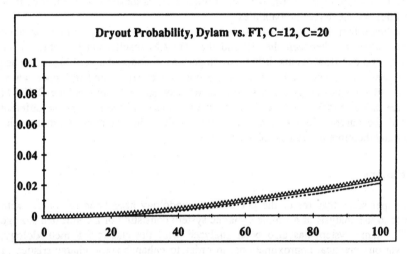

Fig. 5.b. Dryout probabilities with FT (symbols) and with DYLAM (lines) in the case of stochastic failures and overflow level of 12 and 20 meters (dotted).

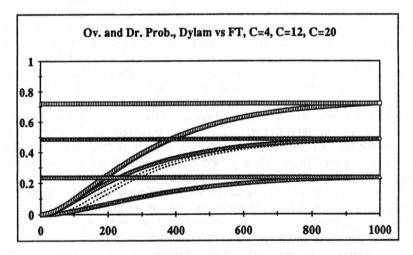

Fig. 6. FT and DYLAM Overflow probabilities for different overflow levels. FT results are presented in symbols: squares refer to the upper bound for the overflow and diamonds represent the correct estimate, triangles indicate the dryout. DYLAM results are indicated by lines: dotted lines represent the overflow for various tank capacities; the dryout cannot be distinguished from the FT estimate.

Finally, figure 6 summarizes all the results discussed above with reference to a much longer time of analysis (1000 hours, equivalent to 1000 time steps). As usual, symbols are associated to the FT results, where diamonds and triangles relate respectively to the overflow and dryout events, while squares represent the upper bound of a FT analysis for the overflow probability. Horizontal lines characterize the probabilities for the lumped failures approximation in the FT approach (see section 4.2.1). The DYLAM results are reported for the three different overflow levels considered (C=3, C=12, C=20), with the dotted lines associated with the overflow probabilities and the dryout probabilities.

It appears that all the dryout estimates are very close to each other, with no significant difference between the FT and the DYLAM results. On the other hand, concerning the overflow events, it is firstly important to notice the very big difference existing between the correct FT value, corresponding to eq. 5, and its upper bound estimate which employs only the first term at RHS of eq. 5. Moreover, the DYLAM results are rather sensitive to the capacity of the tank, as can be seen from the different shapes of the curves. Obviously the asymptotic values do not depend on the tank capacity and therefore the values coincide.

5 Conclusions

In this paper some features of the DYLAM methodology have been presented, with reference to the analysis of a simple dynamic system consisting of a controlled storage tank. The same system has also been analyzed with the classic FT methodology, within various practical approximations, in order to enhance some characteristics of the dynamic analysis that DYLAM permits. The results reported show clearly that, increasing of the capacity of the tank, i.e. increasing the time constant of the system, there is a significant difference between the FT and the DYLAM results, while for a reduced capacity of the tank, the difference can be masked by the approximations contained in the DYLAM code. As a general comment, we can say that, dynamic approaches, employing the physical modeling of the system, are needed when the deterministic and stochastic behaviors of the system are strongly coupled, e.g. as in the case of dependency of the reliability data form the process variables and/or in the presence of complex control loops and man machine interaction.

The DYLAM-3 code offers a general purpose and structured approach for tackling these problems in a comprehensive way. Concerning the capabilities of DYLAM, it is shown here that the code can handle stochastic events, such as random failures. The drawbacks are the need to provide a numerical simulation of the system and the running times of analysis that can be very high, even for simple systems, and much higher than other approaches adopting ad-hoc solutions directly tailored on a specific application. Since DYLAM retains an event tree structure, its practical capabilities of handling stochastic events is related to the number of sequences to be considered, which is a function of the number of components, of their states and of the number of time steps of analysis. The number of process variables that can be treated does not constitute a limitation. Dynamic methods should concentrate on a detailed analysis of short incidental sequences and therefore a reduced number of components with stochastic transitions could be of interest. However, it is our opinion that a dynamic methodology like DYLAM, which retains the event tree structure, can be very suitable in order to generate dynamic event trees, with a finite number of possible branches, where the time of interventions of components and/or of human actions are triggered by the simulation itself (Cojazzi et al. 1992, Cacciabue et al. 1992b). For exploring and quantifying stochastic scenarios with many components failing and repairing, a Monte Carlo approach which generates from scratch a predefined number of independent sequences looks more promising.

References

Acosta C.G., Siu N. (1991) Dynamic Event Tree Analysis Method (DETAM) for Accident Sequence Analysis, MITNE-295, Massachusetts Institute of Technology.

Aldemir T. (1987) Computer Assisted Markov Failure Modeling of Process Control Systems, *IEEE Trans. on Reliability*, R-36 No. 1, 133.

Aldemir T. (1989) Quantifying Setpoint Drift Effects in the Failure Analysis of Process Control Systems, *Rel. Eng. and System Safety*, 24, 33-35.

Belhadj M., Hassan M., Aldemir T., (1992) On the Need for Dynamic Methodologies in Risk and Reliability Studies, *Rel. Eng. and System Safety*, 38, 219-236.

Amendola A., Reina G. (1981) Event Sequence and Consequence Spectrum: a Methodology for Probabilistic Transient Analysis, *Nucl. Sci. Eng.*, 77, 297-315.

Cacciabue P. C., Amendola A., Cojazzi G. (1986) Dynamic Logical Analytical Methodology Versus Fault Tree : the Case of the Auxiliary Feedwater System of a Nuclear Power Plant, *Nucl. Tech.*, 74 (2), 195.

Cacciabue P.C., Carpignano A., Cojazzi G. (1991) Dynamic Reliability Analysis by DYLAM Methodology: the Case Study of the CVCS of a PWR, Proc. Conf. *Probabilistic Safety Assessment and Management (PSAM)*, Beverly Hills, CA, February 4-7, 1991.

Cacciabue P.C., Carpignano A. and Vivalda C. (1992a) Expanding the Scope of DYLAM Methodology to Study the Dynamic Reliability of Complex Systems: the Case of Chemical and Volume Control in Nuclear Power Plants, *Rel. Eng. and System Safety*, 36, 127-136.

Cacciabue P.C., Cojazzi G., Hollangel E., Mancini S. (1992b) Analysis and Modelling of Pilot-Airplane Interaction by an Integrated Simulation Approach, Proc. *5th IFAC/IFIP/IFORS/IEA Symposium on Analysis, Design and Evaluation of Man-Machine Systems*, The Hague, The Netherlands, 9-11 June 1992.

Cojazzi G., Izquierdo J.M., Melendez E. and Sanchez Perea M. (1992) The Reliability and Safety Assessment of Protection Systems by the use of Dynamic Event Trees. The DYLAM-TRETA package, Presented at the *XVIII Reunion Anual de la Sociedad Nuclear Española*, Puerto de Santa Maria-Jerez, October, 1992. CEC-JRC-Technical Note, No. I.92.111.

Cojazzi G., Cacciabue P.C., Parisi P. (1993) DYLAM-3, A Dynamic Methodology for Reliability Analysis and Consequences Evaluation in Industrial Plants, Theory and How to Use. CEC-JRC, Euratom Report, EUR 15265 EN.

Deoss D.L., Siu N. (1989) A Simulation model for Dynamic System Availability Analysis, MITNE-287, Massachusetts Institute of Technology.

Devooght J., Smidts C. (1992) Probabilistic Reactor Dynamics-I: The Theory of Continuous Event Trees, *Nucl. Sci. and Eng.*, 111, 229-240.

Smidts C., Devooght J. (1992) Probabilistic Reactor Dynamics-II. A Monte Carlo Study of a Fast Reactor Transient, *Nucl. Sci. and Eng.*, 111, 241-256.

Limnios N., Cocozza-Thivent C. (1992) Reliability Modelling of Uncontrolled and Controlled Reservoirs, *Rel. Eng. and System Safety*, 35, 201-208.

Marseguerra M., Zio E. (1992) Approaching Dynamic Reliability by Monte Carlo Simulation. In T. Aldemir et al. (eds.) *Reliability and Safety Assessment of Dynamic Process Systems*. NATO ASI series F, Vol. 120, Berlin: Springer-Verlag (this volume).

Siu N., (1992) Risk Assessment for Dynamic Systems: an Overview, submitted for publication in *Rel. Eng and System Safety*.

Markovian Reliability Analysis of Dynamic Systems

Ioannis A. Papazoglou

Institute of Nuclear Technology Radiation Protection
National Center for Scientific Research "DEMOKRITOS"
Aghia Paraskevi, 153-10, Greece

Abstract. Application of Markovian models in the reliability analysis of dynamic systems is presented. In the context of the paper dynamic systems are systems whose stochastic behavior is changing with time. The reliability of control/safety systems coupled with the dynamics of protected or controlled systems is also analyzed. Markovian models are compared with static logic models like fault/event trees.

Keywords. Markovian reliability, Dynamic systems, Safety systems, Control systems, Failure probability, Accident probability, Probabilistic safety analysis.

1 Introduction

The objective of this paper is to present the application of Markov models in the reliability analysis of dynamic systems. Of particular interest to this work are systems whose the stochastic behavior is changing with time. It is this particular dynamic behavior that necessitates the use of techniques more powerful than traditional static models like fault trees, event trees, block diagrams and so on. It is argued that Markov random processes are especially suitable for the simulation of systems whose stochastic behavior depends on the state of the system. Furthermore, since the state of the system changes randomly with time the state dependence is transformed to a random dynamic change in the stochastic properties of the system. Classical models like those mentioned above can not accurately model such dynamic stochastic behavior and hence they can not be used to correctly calculate various reliability performance indices. A class of systems characterized by dynamic stochastic behavior and analyzed in this paper is the class of safety systems. Safety systems are systems designed and engineered to monitor, control, protect, and mitigate the consequences of deviations from the normal operating range of other systems like process plants, nuclear power plants, aircraft, computer systems and so on. This paper demonstrates the applicability of Markov models for the reliability analysis of these systems.

The paper is organized as follows: Section 2 presents the fundamental principles of Markovian Reliability Analysis (MRA). Section 3 lists the categories of systems requiring dynamic modelling and demonstrates the differences of Markov models and static logic models in the case of systems with standby redundancy. Section 4 examines the class of standby safety systems, provides a careful definition of the probability of failure of these systems and presents a simple example of the differences of Markov and static models. Section 5 extends the concept of the standby safety system into that of protection systems; reliability of control systems and the dynamics of the protected systems are included; differences between Markov and other approaches are demonstrated through the analysis of an ammonia storing facility. Finally, section 6 offers some concluding remarks.

2 Basic Principles of Markovian Reliability Analysis

The theory of Markov processes [1,2], the application to reliability analysis[3-9], and methods for Markovian reliability analysis of large systems[10,11] are given in the literature. Here only the necessary definitions are presented.

Markovian reliability analysis is based on probabilistic models that describe the stochastic behavior of systems exhibiting the following properties:

a) the system can be described at any time by specifying its state at that time.

b) the time at which a change from one state to another takes place is an exponentially distributed random variable.

. The discrete - state, continuous time Markov models describe a system that can be in any of a finite number of discrete states that can change at any instant of time.

If $\pi_i(t)$ denote the probability that the system is in state i at time t and $\pi(t)$ is the 1xz row vector with elements $\pi_i(t)$ for i=1,2,..z then $\pi(t)$ satisfies the equation.

$$\dot{\pi}(t) = \pi(t)A \tag{1a}$$

where A is a zxz matrix with elements a_{ij} such that:

$\{a_{ij}dt\}$ = the probability that the system will transit to state j during the interval between t and t+dt given that it is in state i at time t.

A first order difference approximation of this equation results in

$$\pi(n+1) = \pi(n)P \tag{1b}$$

with

$$p_{ij} = \begin{cases} a_{ij}\Delta t & \text{if } i \neq j \\ 1 + a_{ii}\Delta t & \text{if } i = j \end{cases}$$

where $a_{ii} = -\sum_{i \neq j} a_{ij}$ and n denotes time $t_n = n\Delta t$.

Equation (1b) can be solved if the states of the system are defined, the elements of matrix **P** are known, and the initial conditions ($\pi(0)$) are determined.

The states of the system are generated from all possible combinations of the components states. If the system consists of N components and the *ith* component has m_i states, then the number of system states, z, is equal to

$$z = \prod_{i=1}^{N} m_i \tag{2}$$

Next the set of possible states, z, is partitioned into to subsets:

a) An *Available* subset S, consisting of states where the system is available to operate successfully; and

b) An *Unavailable* subset , F, consisting of states where the system is unavailable for successful operation.

If the state equation (1b) is solved, then the probability that at a given instant of time (n) the system will be unavailable to perform its function (*Unavailability*) is given by

$$U(n) = \sum_{i \in F} \pi_i(n) \tag{3}$$

3 The Need For Markovian Reliability Analysis

Reliability analyses begin with the development of a logic model for the system such as event trees, fault trees, cause consequence graphs, block diagrams and so on. What these models primarily achieve is a presentation of the logical dependence of the functionability of the system on the constituent parts. In terms of the definitions used in the previous section these logic models provide a mapping between the space of the possible combinations of states of the components and the subsets S - available- and F -unavailable of system-states. Since the state of the system is determined by the state of its components, the probability that the system will occupy a particular state j is

$$\pi_j(n) \equiv \Pr\{S(n) = j\} = \Pr\left\{\bigcup_{i=1}^{N}(S_i(n) = r_{ij})\right\} \tag{4}$$

where $(S_i(n) = r_{ij})$ denotes the event that the ith component occupies state r_{ij}.

Logic models like fault and event trees can provide the solution of eq. (4) and eventually of eq. (3) efficiently and accurately only if the stochastic behavior of the components is either independent of time or independent of each other. If, however, there are *interdependences* among the components that is, if the stochastic behavior of the components depends on the state of other components or in general on the state of the system, then more powerful techniques as Markovian reliability analysis are in order.

From a reliability point of view all systems are dynamic in the sense that they change their state with time. Furthermore this change of state occurs randomly with time since it involves the change of state of one or more components which also happens randomly in time. It then follows that if the system consists of components exhibiting a stochastic dependence on the state of the system, their stochastic behavior changes with time and in particular at random instances of time. It is this dynamic and stochastic behavior of the system that requires a Markovian model.

Particular circumstances generating dependences on the state of the system include:

i. *Standby redundancy.* Standby failure and repair rates are in most cases different than the corresponding on-line rates. If the switching from the standby to the on-line mode of a component depends on the state of the system (e.g. on whether another component is operating or not) then a Markov model is necessary to correctly account for these differences.

ii. *Common extreme environment.* The failure and repair rates of components change significantly under extreme environments. When the occurrence of such extreme environment exhibits a stochastic behavior then the environment can be considered as part of the system and the failure and/or repair rates of the various components depend on the state of the system.

iii. *Components sharing common loads.* If a load is shared by several components, and the failure rates of the components depend on the size of the load, the failure rates of the components depend on the number of operating components that is on the state of the system.

iv. *Systems with repairable components.* This a particularly common situation where dependences on the state of the system arise owing to physical constraints and/or repair policies. Repair of a component may be possible only if the system is operating and not if it has failed catastrophically. Alternatively, repair might be possible only if the system is shutdown and not operating; in this case several repair policies might be possible as "repair all components and then resume operation" or "resume operation as soon as possible". Finally repair might depend on the number of available repairmen.

v. *A class of dependent failures.* In several instances the failure rate of a component depends on the state of another component because failure of the latter might either increase the load (case iii) or generate an extreme environment (case ii). Furthermore, the combined effect of two or more component failures on the success or failure of a system might depend on the order in which the components failed.

vi. *Failure probability of standby systems.* Standby safety systems are systems that must operate when called upon. Correct calculation of the failure probability of such systems require the employment of Markov models.

vii. *Accident probability of protection systems:* This is an extension of the previous case where the dynamic behavior of the protection system is included in the analysis in order to take into consideration the time period between the onset of an accident and the point of no return. During this period it might be possible to stop and reverse the course of an accident.

Specific examples along with quantitative evaluations of the differences between Markov models and static models like fault trees and event trees for cases i-vi are given in Ref [12,13]. Here we will present examples of cases (i), (vi), and (vii).

3.1 Systems with Standby Redundancy

Let us consider a two-component parallel system where one of the components is operating while the other is on warm standby.
The state transition diagram for this system is given in Fig. 1. It is assumed that at time zero component A is operating. It is noteworthy that a transition from state 1 to state 3 and a transition from state 2 to state 4 imply the same component state transition, namely the failure of component B. These transitions are not characterized by the same transition rate, however, since the first occurs while component B is in a standby mode (A is operating) and the second while component B is operating (A has failed).

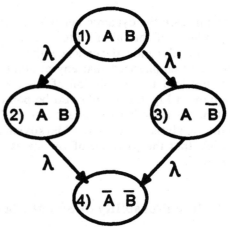

Fig. 1. State transition diagram for a two component system with warm standby redundancy

3.1.1 Markovian Solution

Solution of the state equations implied by the state transition diagram of Fig. 1 yields for the state probability of state 4, which is the unavailability of the system, the following expression

$$U(t) = 1 - \left(1 + \frac{\lambda}{\lambda'}\right)\exp(-\lambda t) + \frac{\lambda}{\lambda'}\exp[-(\lambda + \lambda')t] \tag{5}$$

3.1.2 Static Logic Model Solution

A static model like a fault tree or an event tree would evaluate the failure probability of the system as

$$U(t) = \Pr\{\overline{A} \cdot \overline{B}\} = \Pr\{\overline{A}\} \cdot \Pr\{\overline{B} / \overline{A}\} \tag{6}$$

which can only bracket the correct solution in eq. (5) between

$$U_1(t) = 1 - 2\exp(-\lambda t) + \exp(-2\lambda t) \tag{7}$$

if the standby is considered hot $(\lambda' = \lambda)$, and

$$U_2(t) = [1 - \exp(-\lambda t)][1 - \exp(-\lambda' t)] \tag{8}$$

if it is assumed that component B always exhibits a reduced failure rate λ'.

In this case the dynamic behavior of the system consists in the dynamic dependence of the failure rate of component B on the state of the system. It is this dynamic dependence that a Markov model can simulate. It should be emphasized that any attempt to correctly solve eq. (6) will inevitably lead to the solution of the state equations implied by the state diagram in Fig. 1. It should also be pointed out that a logic model for the system (be it fault tree, event tree etc.) is an indispensable part of reliability analysis even for the Markovian approach since it is through such models that the partition of the states into operating or failed subsets is achieved.

4 Failure Probability of Standby Safety Systems

Another instance where Markovian models are necessary is when the failure probability of standby safety systems is to be calculated. Many of the quantitative results of a probabilistic safety analysis are expressed in terms of the frequency of occurrence of undesired events like the melting of the core of a nuclear power plant or the release of a hazardous material from a chemical installation. Almost universally this frequency is calculated as the product of the frequency of occurrence of an initiating event times the "average unavailability" of standby safety systems. It will be shown here that this practice provides only an approximation and not necessarily a good one. Without loss of generality the discussion will be confined into a single standby system. The key concepts of unavailability on demand and failure probability will be defined and the differences will be discussed.

Usually a safety system remains in a standby mode until there is a need for it to operate. An undesirable event (an accident) occurs if the system is not available to operate when challenged to do so. Before the onset of the challenge, however, it can fail and if the failures are detectable it can be repaired. What is of importance here is the coincidence of a demand and the system being unavailable. Hence, the probability that an accident will occur during a time period T is the probability that a challenge will occur at some instant in the period T and at that instant the system is unavailable. Correct calculation of the accident probability requires proper handling of the dependences between the frequency of the demand and the unavailability of the system as follows.

Success probability S(t) of a standby system is the probability that during a predetermined time period (t) there will be no instant for which there is a demand for the system *and* the system is unable to perform its function.

Failure probability F(t) of a standby system is the probability that *at least* once during the predetermined period of time (t) there will be a demand for the system *and* the system will not be able to perform its function. F(t) is complementary to S(t).

Correct calculation of F(t) requires the enlargement of the state-space of the safety system by a subset of states called Failed states (see Fig. 2). The failed states are into one-to-one correspondence with the unavailable states with a significant addition: a demand for the system is present.

If demands are arriving according to a Poisson random process with rate λ_0, the arrival of a demand can not cause a system failure, and the arrival of demands does not depend on whether an accident has occurred, then it can be shown (see [14]) that

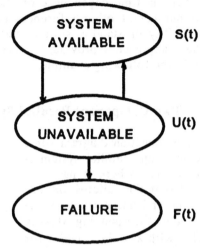

Fig. 2. Schematic transition diagram for standby safety systems

$$F(t) = 1 - \exp\left(-\lambda_0 t \overline{U}(t)\right)$$

$$\text{where } \overline{U}(t) = \frac{1}{t}\int_0^t U(x)dx \tag{9}$$

$\overline{U}(t)$ in eq. (9) is the average unavailability of the system in the period [0,t] *given* no accident in [0,t].

$$\text{If } \lambda_0 t \overline{U}(t) << 1 \quad \text{then (9)} \quad \Rightarrow \quad F(t) = \lambda_0 t \overline{U}(t) \tag{10}$$

It is this latter relationship in eq. (10) that most probabilistic analyses are using to calculate the frequency of an accident conditional on no accident up to time t. Regardless of whether eq. (9) or eq. (10) is used, however, the correct calculation of $\overline{U}(t)$ is required. This means that the average unavailability *given* no accident has occurred up to time t need be calculated. Several models for the unavailability of standby systems have been presented in the literature. Reference [15] provides approximate analytical solutions for 1-out-of-2, 1-out-of-3, and 2 -out -of -3 systems under periodic test and maintenance and different testing policies that correctly account for the time dependence of the phenomena. The same reference sites a number of other references that address various aspects of the unavailability of standby systems under periodic test and maintenance. All these approaches provide, however, only approximate answers as is shown in the following example.

4.1   Two Component Parallel System

The state-transition diagram for a two component parallel system is given in Fig. 1 where it has been assumed that the failures are undetectable and unrepairable. Solution of the corresponding Markov model $\left(\text{with } \lambda' = \lambda\right)$ yields for state 4 (two components down) U(t) as in eq. (7)

This probability is the probability that both components will be down at t and hence that the system will be unavailable. The same result could have been obtained with other methods such as fault trees, reliability block diagrams, or state enumeration techniques. The average unavailability over a period T is then given by

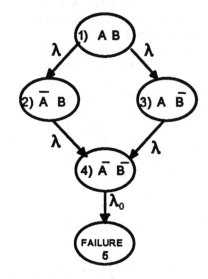

Fig. 3. State transition diagram for a two component parallel system with a failed state

$$\overline{U}(T) = 1 - \frac{2}{\lambda T}\left[1 - \exp(-\lambda T)\right] + \frac{1}{2\lambda T}\left[1 - \exp(-2\lambda T)\right] \qquad (11)$$

This expression, however, can not be used in eq. (9) or eq. (10) to provide the correct answer for the failure probability since it provides the average unavailability of the system independently on whether an accident has occurred or not. To correctly account for the fact that unavailability of the system should be calculated conditionally on the absence of an accident a fifth state should be added as in Fig. 3. This is an absorbing state and the system can not leave it once entered. The probability of occupying it at time T is the failure probability of the standby system that is, the probability that an accident will happen in the interval [0,T]. Solution of the corresponding Markov model yields

$$F(t) = \pi_5(t) = 1 - \frac{2\lambda_0}{\lambda_0 - \lambda} \exp[-\lambda T] + \frac{\lambda_0}{\lambda_0 - 2\lambda} \exp[-2\lambda T] - \frac{2\lambda^2}{(\lambda_0 - 2\lambda)(\lambda_0 - \lambda)} \exp[-\lambda_0 T] \quad (12)$$

The values of the failure probability given by eq. (12) are lower than those obtained when eq. (9) and eq. (11) are combined. This is due to the fact that the model in Fig. 1 allows the system to be exposed to challenges continuously throughout period T regardless of whether a failure has already occurred. This would have been right only if (instantaneous) restoration of the system after the accident was possible and hence   eq. (10) would give the expected number of demands coinciding with the system being unavailable. If, however, the probability that such an event will occur even once in a period is of interest, then the model in Fig. 3 is the right one. In this model the probability of the system being unavailable is, as in the model of Fig. 1, the probability of occupying state 4 . Now, however, $\pi_4(t)$ gives the unavailability of the system given that no accident has occurred.

Reference [14] presents results of the comparison of eq (12) and eq (10) where eq. (11) is used for $\overline{U}(t)$. It is shown that the overestimation of the "conventional approach" ranges from a factor of two to two orders of magnitude depending on the ratio of λ_0/λ.

5. Accident Probability in Control/Safety Systems

Another situation where Markov models constitute a more powerful and necesary analysis tool is the case where protection systems are involved. Similar problems are treated in [16-20].

Protection systems aim at protecting another system, usually one that exhibits a dynamic behavior like a nuclear power plant, a chemical plant, an aircraft etc. In general, they can be divided into two classes:

(a) Protection systems that monitor a set o physical parameters of the protected plant and take appropriate action if any or a combination of them exhit a predetermined desirable region. The function of these systems is similar to that of *control systems* and the different name simply indicates the safety implication of a failure of the particular control systems. These systems are operating continuously (monitoring the parameters) but their failure becomes significant only if it coincides with a deviation of the monitored parameters.

(b) Protection systems that are engineered to provide a specific safety function in the event that a particular combination of other failures brings the vital parameters of the plant outside the acceptable region. These *safety systems* are usually in a standby mode and are required to operate if called upon by the occurence of an abnormal event.

Both these systems can fail, but as explained in the previous section their failure is not critical unless it coincides with a challenge. Failures of protection systems can be either immediatelly detectable with repair starting immediately, or undetectable. In the latter case failures may be detected after tests. In a large class of aplications if a challenge occurs while the protection system is unavailable the change is so rapid when compared with the times necessary to repair the protection system that the occurrence of the accident is assumed instantaneous and the situation is not recoverable. In this case the analysis of the previous section is applicable.

There are instances however, in which the protection system can change its state in time scales comparable to those of the physical processes that follow the onset of a challenge. This is usually the case when some components of the system can change their state through operator's action. Accidents do not happen instantaneously with the onset of a deviation from normal operation conditions. It takes some time for the appropriate variables (temperature, pressure, etc.) to reach the limits beyond which a catastrophic failure occurs. It might be possible that even if a protection system was in a failed state at the onset of an accident, it can transit back to an operating state before it is too late. This delay or grace period from the onset of an accident to the final catastrophic failure depends on the initial physical conditions and the failure mode of the systems. The dynamics of the safety systems interacts with the evolution of the physical process and therefore the delay may increase or decrease during the accident as the protection and the protected systems change their state. The probability of an accident depends, therefore, on the delay times as well as on the failure and repair rates and in this case both dynamic phenomena need be modeled.

In this case the state space must be enriched by considering the states

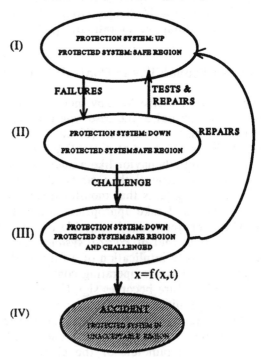

Fig.4. State transition diagram of protection system

of the protected system. The enlarged sate space can be now partitioned in four subsets or regions as follows:(see also Fig. 4)

Region I: States where the protection system is available and the relevant parameters of the protected system are in the safe region.

Region II: States where the protection system is unavailable and the relevent parameters of the protected system are in the safe region. (No challenge present).

Region III: States where the protection system is unavailable, the relevent parameters of the protected system are in the safe region but there is a challenge to the system. Starting from such a state the parameters are drifting towards the unacceptable region.

Region IV: States where the parameters of the protected plant have entered the unacceptable (accident) region.

Failures, tests and repairs of components of the protection system while no challenge is present correspond to transitions within region (I), from (I) to (II), within (II), and from (II) back to (I). A challenge occuring while in region (I) has no particular efect other than revealing any partial failures in the protection system, continuation of the operation of the plant or safe shutdown. A challenge occurring while in region (II), however, will result (unless somehow mitigated) into the plant's parametrs entering the accident region. For each state in region (III) the time required for the physical processes to bring the plant into the accident region can be determined and may vary from state to state. Repairs of the protection system might, however, be possible. Such repairs will either bring the protection system back to one of its available states or will constitute a partial repair and the combination of protection and protected systems will remain in a state of region (III). In the latter case a new competitive process takes place between protection system repair and plant parameters entering the unacceptable range. This continues until the system either returns in region (I) or enters region (IV).

This process can be accurately simulated by a Markov model as follows.

The state-space is enlarged by a number of state variables that determine the state of the physical process. If the system has N components each with m_i states, and K process variables each with k_j states, then the total number of system states, z, is equal to

$$z = \prod_{i=1}^{N} m_i \prod_{j=1}^{K} k_j \tag{13}$$

In other words a generalized state-space has been created resulting from the Cartesian product of the control system state-space and the process variable state-space.

Let:

i, j denote two states of the generalized state-space

c_i, v_i denote the control system state and the process state that correspond to the generalized state i

c_j, v_j the corresponding quantities for the generalized state j

p_{ij} (see eq.(1)) is then determined as follows:

First it is determined whether a transition between process states v_i and v_j is possible (as it is described later). If this is possible p_{ij} is set equal to the transition probability from c_i to c_j If a transition from v_i to v_j is not possible then p_{ij} is equal to zero.

5.1 Process Change of State

Let $\mathbf{x} = (x_1, ..., x_k)$ denote the vector of the process state variables. Let $f(\mathbf{x})=0$ denote the equation of state change of the process. This equation will in general depend on the state c, of the control system, so it can be written as

$$f_c(\mathbf{x}) = 0 \tag{14}$$

The system transits among the system states (due to failures and repairs of components) at times which are randomly distributed. This stochastic behavior is thus transferred to the change of state of the process (through eq.(14))

If eq. (14) is approximated by a first order differential equation

$$\dot{\mathbf{x}} = g_c(\mathbf{x}) \tag{15}$$

and this is in turn approximated by a first order difference equation, the following state-equation for the process state results:

$$\mathbf{x}(n+1) = \mathbf{x}(n) + \Delta t g_c[\mathbf{x}(n)] \tag{16}$$

Given a control system state c_i and a process state v_i [defined by $\mathbf{x}(n)$] it follows from eq. (16) that the process state at time n+1 is uniquely determined as $\mathbf{x}(n+1)$ ($\equiv v^*$). This in turn means that.

$$\Pr\{(c_i, v_i) \to (c_j, v_j)\} = \begin{cases} 0 & \text{if} \quad v_j \neq v^* \\ c_{ij} & \text{if} \quad v_j = v^* \end{cases} \tag{17}$$

Denoting now the four state probability vectors corresponding to the four state regions by $\left[\pi_I(n), \pi_{II}(n), \pi_{III}(n), \pi_{IV}(n)\right]$, results in the following form for the state equations:

$$\pi(n+1)=[\pi_I(n),\pi_{II}(n),\pi_{III}(n),\pi_{IV}(n)]\begin{bmatrix} P_{I,I} & P_{I,II} & 0 & 0 \\ P_{II,I} & P_{II,II} & P_{II,III} & 0 \\ P_{III,I} & 0 & P_{III,III} & P_{III,IV} \\ 0 & 0 & 0 & I \end{bmatrix} \quad (18)$$

Solution of eq. (18) provides the accident probability for the interval (0,n) as $\pi_{IV}(n)$.

The dependence of the stochastic behavior of the protection system on the state of the system is introduced through the physical process described by eq. (15) and the random times at which the components of the protection system change states. An event/fault tree approach (provided that there are no other dependences) would bracket the answer of the Markov model, by overestimating the accident probability if no repair during the grace period is assumed and underestimating the accident probability if even partial recoveries are assumed successful.

5.2 An Example

Consider a system consisting of a tank storing ammonia, its associated refrigeration and control system, and a pipe section connecting the tank with an ammonia consumption site (See Fig. 5). Ammonia is stored as a refrigerated liquid at atmospheric pressure conditions (i.e. -33^{0}C and 1 bar)

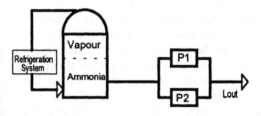

Fig. 5. Simple diagram of the tank

5.2.1 Event Tree/Fault Tree Approach

An important initiating event (IE) that can cause an accident is the failure of the refrigeration system. The event tree corresponding to this initiator is given in (Fig. 6).

A reduction in the capacity of the refrigeration system during storage initiates a transient and in order to avoid release of ammonia discharge of the

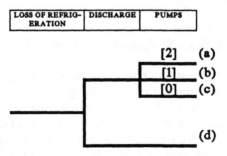

Fig. 6. Event tree: loss of refrigeration during storage

contents of the tank is required (safety function) . The event Discharge (Fig. 6) models the availability of necessary hardware (pipes, valves excluding pumps) for unloading the ammonia of the tank. If discharge is not possible, then there will be an accident(sequence (d) in Fig. 6). Otherwise, the number of available pumps and the initial values of the Pressure and Liquid level in the tank determine the occurrence or not of the accident.

In the event tree - fault tree analysis, event sequences (c) and (d) lead to an accident because in both cases discharge is not possible either because of pumps unavailability (c), or other hardware unavailability (d). Event sequence (a) doesn't lead to an accident, because if two pumps start to operate when the refrigeration system fails the tank will be emptied before the pressure rises to an undesirable level. Accident sequence (b) leads to an accident, because one pump is not enough to empty the tank before pressure rises beyond the accident level. In the fault tree analysis the probabilities of the accident sequences are estimated as follows:

sequence b: $P_{rf} * P_{p1}$

sequence c: $P_{rf} * P_{p2}$

sequence d: $P_{rf} * P_d$

where: P_{rf} is the frequency of failure of refrigeration system

P_d is the unavailability of the discharge system

P_{p1} is the unavailability of one pump

P_{p2} is the unavailability both pumps

Recovery, that is repair of the failed hardware, can be included as follows: A grace period is defined for each accident sequence. This time period is defined as the time necessary for the process variables, starting from the steady state value, to reach their unacceptable values. In this case the pressure of ammonia should not exceed an unacceptable limit. A probability of recovery is then defined as the probability to restore the failed equipment within the grace period. The complementary probability of non-recovery can finally be used to multiply the frequency of each accident sequence. The frequency of the three accident sequences then becomes.

sequence b: $P_{rf} * P_{p1} * P_{nrb}$

sequence c: $P_{rf} * P_{p2} * P_{nrc}$ (19)

sequence d: $P_{rf} * P_d * P_{nrd}$

where: $P_{nrb}, P_{nrc},$ and P_{nrd} are the probabilities that there will be no repair.

This of course is an approximation stemming on one hand from the inability of determining exactly the grace period (since it depends on the initial conditions of the process) and on the other from the fact that such repairs usually mean transfer from one accident sequence to the other. For example, repair of one pump in accident sequence (c) means transfer to the accident sequence (b), while repair of both pumps means transfer to the safe sequence (a).

Consequently one might either underestimate the accident probability by assuming successful response even if partial recovery is achieved, or overestimate the accident probability by ignoring recovery or through other conservative assumptions about the available grace periods.

Accordingly two sets of results have been calculated: one assuming that no recovery is possible and one assuming that if any of the failed components (in an accident sequence) is repaired within the grace period for this sequence, then the sequence is successfully terminated. The results are discussed in the following subsection.

5.2.2 Markovian Reliability Analysis

Markovian analysis starts with the generation of system states. The system states are combinations of all possible states of the components and of all possible states of the process variables. The components in this case are the refrigeration system, the pumping system and the discharge system. The process variables are the tank pressure and the tank liquid level. The refrigeration system can be in two states. In state No 1 the refrigeration system is operating and in state No 2 it has failed. The discharge system can be in three states. In state No 1 the discharge system is operating, in state No 2 it has failed but the failure is undetected and in state No 3 it is under repair. The pumping system can be in 5 states. In state No 1 both pumps are operating, in state No 2 one pump is operating and the other has failed but its failure is undetected, in state No 3 both pumps have failed and their failure is undetected, in state No 4 one pump is operating and the other under repair, in state No 5 both pumps are failed but the failures are detected and they are under repair. Transition to states involving detection of failures occur when the process variables reach the level that trigger the corresponding alarms.

The pressure and the liquid level of the tank are discretized into several states between the maximum and minimum values they can take. The pressure of the tank can take the values between 1 and 1.2 atm. The number of states of the physical variables pressure and level depend on the time step of the solution of eq. (15). In this application 40 and 64 states have been considered for the pressure and the level, respectively.

Seven groups of states can be distinguished on the basis of the state of the control/safety system.

(a) States in which the refrigeration system is operating and the discharge subsystem along with one or two pumps are available.

(b). States in which the refrigeration system is failed, the discharge subsystem and both pumps are operating.

Groups (a) and (b) form region I of Fig. 4.

(c). States in which the refrigeration system is operating, the discharge subsystem is down and/or both pumps are unavailable. This group corresponds to region II in Fig. 4.

(d). States in which the refrigeration system is failed, the discharge subsystem is operating and one pump is operating.

(e). States in which the refrigeration system is failed, the discharge system is operating and both pumps are failed.

(f). States in which the refrigeration system is failed and the discharge system is failed.

Groups (d), (e) and (f) when combined with physical parameters in the acceptable region (i.e. pressure below 1.2 atm) form region III in Fig. 4. The same groups of states when combined with physical parameters in the unacceptable region form the accident superstate corresponding to region IV in Fig. 4.

Reference [21] describes the development of the state equations according to eq. (18). Three sets of calculations have been performed and the results are shown in Fig. 7.

The curve marked [ET/FT] shows the resulting failure probability if no recovery is considered in the ET/FT approach. The curve marked "ET/FT with repair" shows the resulting failure probability in the ET/FT approach if recovery is considered as follows. Starting with the initial (steady state) conditions, the process state equations are solved for each group of accident sequences [(b),(c),(d)]. From the resulting solution the time required for the pressure to exceed the predetermined safety limit (1.2 atm) is determined. These times t_b, t_c, t_d provide the grace periods for each accident group. Next, the probability of repairing at least one of the failed components in the accident sequence within the corresponding grace period is calculated. This provides the recovery probability and its complementary the nonrecovery probability used in eq.(19).

The two curves are bracketing the exact answer but they represent a band of three orders of magnitude.

Finally the curve marked Markovian provides the more realistic results of the Markovian model. The results show that for the set of parameters values considered in this example the ET/FT model with recovery underestimates the failure probability by a factor of 5 while if the model without recovery is used the failure probability would have been overestimated by two orders of magnitude.

6 Concluding Remarks

The need for and the applicability of Markovian Reliability Analysis (MRA) has been demonstrated for a number of classes of dynamic systems. MRA is necessary when the stochastic behavior of the system (that is of its components) exhibits a dependence on the state of the system and hence changes randomly with time. Such dependences arise in a number of cases including systems with components in cold or warm standby, components sharing common loads or with environmentally dependant stochastic characteristics, and in various types of repairable systems.

MRA is particularly useful in calculating the probability of accidents where a challenge to a safety system to perform its safety function must coincide with a failure of the safety system.

Markovian models offer the extra capability of simulating various quantitative measures of reliability and safety performance but they can not replace logic models providing the logical interconnections of the components in a system. Furthermore MRA poses formidable numerical problems since the number of states in a system increases exponentially with the number of the components. This problem becomes more severe if physical processes must be modelled along with failures and repairs of components. Improved algorithms of solving the state equations [1] along with ever increasing storage and speed capabilities of the computers helped alleviating this problem[10,11]. It is the opinion of this author that numerical complexity is not a real problem any more. A final hurdle to the wider use of MRA is the user-friendlieness of the various computer programs. Definition of a system and generation of the transition probability matrix incorporating the various dependences need be automated in a truly user-friendly environment before MRA finds the wide applicability it deserves.

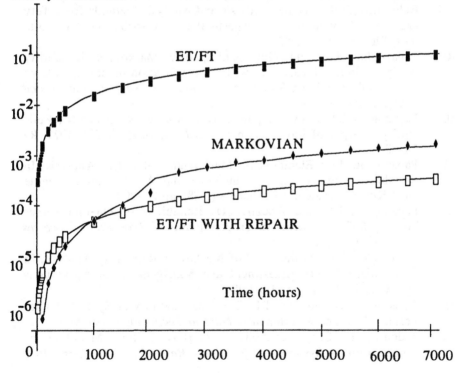

Fig. 7.   Failure Probability versus Time.

References

1. Howard, R.: Dynamic Probabilistic Systems. Vols. I & II, Wiley (1971)
2. Kemeny, J.G. and Snell, J.L.: Finite Markov Chains. D. Van Nostrand (1961)
3. Lees, F.: Loss Prevention in the Process Industries. Vol. 1, Butterworths, Guilford, UK, 1986, Chapter 7
4. Henley, E. and Kumamoto, H.: Reliability Engineering and Risk Assessment. Prentice-Hall (1981)
5. Shooman, M.D.: Probabilistic Reliability: An Engineering Approach. McGraw-Hill (1969)
6. Singh, C. and Billinton, R.: System Reliability Modeling and Evaluation. Hutchinson, London (1977)
7. Dhillon, B.S. and Singh, C.: Engineering Reliability: New Techniques and Applications. Wiley (1981)
8. Endernyi, J.: Reliability Modeling in Electric Power Systems. Wiley (1978)
9. Blin, A., Carnino, A. and Georgin, J.P.: Use of Markov Processes for Reliability Problems. In Synthesis and Analysis Methods for Safety and Reliability Studies. G. Apostolakis, S. Garriba, and G. Volta (eds) Plenum Press (1980)
10. Papazoglou, I.A. and Gyftopoulos, E.P.: Markovian Reliability Analysis Under Uncertainty with an Application on the Shutdown System of the Clinch River Breeder Reactor. *Nuclear Science and Engineering* 1980, 73,1
11. Papazoglou, I.A. and Gyftopoulos, E.P.: Markov Processes for Reliability Analysis of Large Systems. *IEEE Trans. Reliability* 1977, R-26, 232
12. Papazoglou, I.A.: Elements of Markovian Reliability Analysis. In Reliability Engineering. A. Amendola and A. Saiz deBustamante (eds) Kluwer Academic Publishers (1988), pp 171-204
13. Papazoglou, I.A. and Aneziris, O.: Reliability of On-line versus Standby Safety Systems in Process Plants. *J. Loss Prev. Processes Ind.* 1990, Vol 3, 212
14. Papazoglou, I.A.: On the Need of Markovian Reliability Analysis. In Probabilistic Safety Assessment and Management. G. Apostolakis (ed.) Elsevier (1991), pp 1413-1418
15. Apostolakis, G. and Chu, T.L.: The Unavailability of Systems Under Periodic Test and Maintenance. *Nuclear Technology* 1980, 50,5
16. Aldemir, T.: Computer-Assisted Markov Failure Modeling of Process Control Systems. *IEEE Trans. Reliability* 1987, R-36, No 1, 133

17. Aldemir T.: Utilization of the Cell Mapping Technique to Construct Markov Failure Modes for Process Control Systems. In Probabilistic Safety Assessment and Management. G. Apostolakis (ed) Elsevier (1991)

18. Devooght, J., Smidts, C.: Probabilistic Reactor Dynamics I. The theory of continuous event trees. *Nuclear Science and Engineering* 1992

19. Smidts, C., Devooght, J.: Probabilistic Reactor Dynamics II A Monte Carlo study of a fast reactor transient. *Nuclear Science and Engineering* 1992

20. Devooght, J., Smidts, C.: Probabilistic Reactor Dynamics III. A framework for time dependent interaction between operator and reactor during a transient involving human errors. *Nuclear Science and Engineering* 1992.

21. Papazoglou, I. A., Anezirs, O.: Reliability of Control Systems in Chemical Plants. Proc. ESREL-93, Munich, 1993

Approaching Dynamic Reliability
By Monte Carlo Simulation

Marzio Marseguerra, Enrico Zio

Dept. of Nuclear Engineering – Polytechnic of Milan, Milano, Italy

Abstract. The event tree/fault tree methodology is probably the most widely recognized and adopted method for assessing the reliabilty characteristics of a plant. However the methodology might suffer some limitations when treating scenarios in which the dynamics of the system play an important role. In these cases a dynamic approach seems appropriate: such analysis should take into account the mutual effects of the process physical evolution, the operator's performance and the system hardware configuration. In general this analysis requires large efforts both in the implementation of the relating models and in their evaluation as a function of time, which in turn leads to large CPU times. Therefore, a research effort aiming at developing fast models and more efficient techniques for their evaluation is needed. Moreover the effects of the input uncertainties on the resulting output should be considered, this being particularly true for the Human Reliability Analysis (HRA) both for the operator model and its input data. In this regard the Monte Carlo approach seems to enjoy the desirable features.

In this paper we present a Monte Carlo dynamic approach to reliability and compare it to a classical static analysis in which the time–dependent response of the plant and the control are not followed. A simple example shows how in those cases in which the reliability characteristics of the plant strongly depend on the process, the results obtained by the two methodologies are quite different. The simulation allows to take into account some important issues such as components' aging, failures on demand, operator's actions. Here we present a simple model of failures on demand and the effects that these failures have on the reliability of the system. Biasing of such failures is also introduced and in this regard we point out that these techniques in general require a high degree of care and judgment.

Keywords: dynamic reliability, Monte Carlo simulation, variance reduction techniques, protection–control systems, failure on demand.

1 Introduction

A classical method for assessing the reliability of a real plant is the event tree/fault tree (ET/FT) methodology. Starting from an initiating event, the event tree details the possible sequences of events in terms of success or failure of the systems designed to mitigate the effects of the accident. The fault tree is used to evaluate the success and failure probabilities of each system in terms of those of its components. In this scenario, however, it is very important to note that when a system changes its configuration, i.e. when one of its components makes a state transition, variations in one or more of the process variables may also originate. If these variations do not appear or their effect is negligible, then the ET/FT methodology leads to satisfactory results. On the contrary, if the system behaviour is strongly sensitive to the process variables values, then any reliability analysis should properly take into account these dynamic aspects and therefore a dynamic PRA approach is almost mandatory. Unfortunately this approach introduces a great deal of difficulties, essentially due to the fact that while the classical methods may be carried out in terms only of the hardware stochastic characteristics, now, for each hardware configuration, a suitable model for the process variables evolution must be introduced. Notice that a real plant is likely to have a large number of configurations and accordingly a large number of physical models is needed. In addition, while the stochastic transitions occur on the large time scale of the component stochastic time constants, the physical processes have a much quicker evolution and thus need to be followed with suitably small time steps, leading to a large use of CPU time. To be as practical and realistic as possible, this difficulty may be faced only by using reasonably simplified physical models and/or implementing efficient techniques for their solution, such as the Response Surface Method (RSM) [1] and the Cell–to–Cell Mapping Technique (CCMT) [2–4]. In our opinion, another possibility worthwhile to explore is the application of the artificial neural network methodology [5]. The above scenario becomes even worse if one considers that a more realistic plant description requires consideration of the human factors. Here a variety of techniques has been developed but it seems that still more efforts should be devoted in this area. In our opinion, a possible approach might consist in "proceduralizing" the operator's actions thus assimilating the operator to a component with well defined rules. In this respect we are conscious that this "behavioural model" is a good way to skip the problem.

In the present paper we addressed the problem of dynamic PRA within the framework of the Monte Carlo simulation. Since the main features of this method are well known [6–11], here we limit ourselves to pointing out that the event simulation approach enables the analyst to take into account many relevant aspects of the dynamic evolution of a plant. In particular it is possible to consider the aging of the components, the effects of the failures on demand on protection/control systems and, most important, the influence of the process variables on the hardware stochastic characteristics. This last point allows a deep investigation of the ordering and timing of the events occurring in the accident scenario [4], [12].

In Section 2 we enlight some of the above mentioned features of the Monte Carlo approach in dynamic reliability analysis. In Sections 3 and 4 a simple case study [4], [13–15], is described and the results thereby obtained are summarized. In Section 5 a feature of the protection/control system is investigated. The last section is devoted to the concluding remarks.

2 Monte Carlo in Dynamic Reliability

The Monte Carlo method may be briefly described as a technique of studying an artificial stochastic model of a physical or mathematical process. The "calculation" involves playing a game of chance using random sampling techniques. The game is carried out in such a way that the expected value of the score is the desired physical or mathematical quantity. Then, if the game is played enough times, the average scores approach the expected values, i.e. the quantity to be calculated [16–18].

The framework provided by a Monte Carlo simulation may be extended to consider dynamic reliability problems, with allowance given to the interactions between the hardware, the process variables and the protection/control systems.

In analog Monte Carlo [10], given that a system at time t' is in state k', the time t of the next transition is sampled from the probability density function $f(t|k', t')$; then the new state k is sampled from the transition probability $q(k|k', t)$. This procedure is well suited for taking into account the dynamic behaviour of the plant and its control. To do so, physical models which describe the time evolution of the plant in different hardware configurations should be preliminarily prepared. If $y(t)$ is the vector of the involved process variables, the physical model for the system in configuration j at time t is represented by the following vector equation

$$\mathbf{f_j}[\mathbf{y}(t), \dot{\mathbf{y}}(t), t; \mathbf{a_j}, \mathbf{y}(t_j^0)] = 0 \qquad t \geq t_j^0 \tag{1}$$

where $\mathbf{j} \equiv (j(1), j(2), \ldots, j(N_c))$ is the vector identifying the states of the N_c components constituting the system, $\mathbf{y}(t_j^0)$ is the vector of the process variables at time t_j^0 at which the system enters configuration $\mathbf{j}$ and $\mathbf{a_j}$ represents the set of parameters pertaining to that configuration. Once the equations for the physical models have been implemented, it is possible to simulate complete histories comprising stochastic as well as deterministic, "control-imposed" transitions. The general procedure is illustrated in Figure 1. At the beginning of the simulation and each time a stochastic or a control–imposed transition occurs, the time of the next stochastic transition and that of the next control intervention are computed: the first occurring event determines how the system configuration must be updated. The figure refers to a Monte Carlo trial in which the first occurring event is the stochastic transition at time t_1. This means that during $(0, t_1)$ the process variables, which evolve according to the physical model PM1, do not exceed limits suitably specified by the protection/control systems. Once the transition has been selected by random sampling from the involved distributions, the system changes accordingly and its evolution now follows the physical model PM2. At t_1 the time t_2 of the next stochastic transition is also sampled. In the case shown in the figure, the system physical evolution is such that a protection/control intervention is demanded at $t^* < t_2$. At this point the trial continues according to one of the two following possibilities: if a failure on the demand occurs, then the system continues to evolve following the previous physical model PM2, and the time of the next stochastic transition remains at t_2. In the opposite case, the protection/control intervention modifies the system configuration so to keep the process variables within the pre–established limits. In this new configuration the system evolves according to the physical model PM3. In addition, the time t_3 of the successive stochastic transition is sampled. The described sequence is carried on until the system fails or the mission time T_{miss} is reached. Notice that the dynamic approach introduces new and different definitions of failure criteria: now a failure is also generated when the process variables exceed pre–established threshold values. In other words, when a process variable, e.g. a temperature, exceeds a certain threshold, the system is said to have reached a top event condition [19]. As mentioned in the Introduction, the efficient solution of eqs.(1) represents a crucial point for the practical feasibility of a dynamic reliability analysis. In this regard we are now exploring the potential of the artificial neural network methodology in solving suitably discretized differential equations [5]. Here we limit ourselves to mention an example concerning the 12–steps ahead solution of a non–stationary problem described by a given ARMA(13,13) model. In Figure 2 the solution obtained by the neural network and that computed by the model are compared and the good agreement looks promising.

In the next section we shall present a simple application of dynamic Monte Carlo and compare it to the classical reliability analysis.

3 A Case Study: The Holdup Tank

The Monte Carlo simulation approach has been applied to a simple problem taken from literature [4], [13–15]. The system consists of a tank containing a fluid whose level is determined by three components (Figure 3).

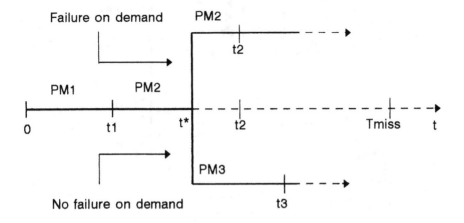

Figure 1 Example of two possible sequences in a Monte Carlo trial

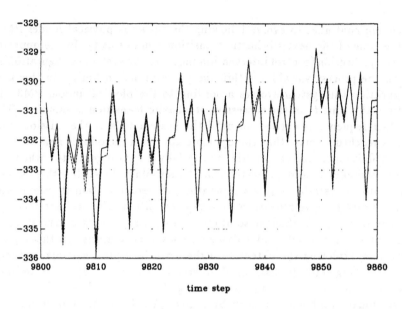

Figure 2 Example of a 12–steps ahead solution of a non–stationary problem described by a given ARMA(13,13) model. Solid line represents the ARMA model values; dashed line represents the neural network solution.

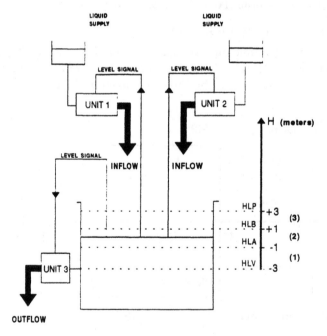

Figure 3 Diagram of the holdup tank.

The components work independently and each of them has four possible states, namely: 1 = operating (ON), 2 = stand-by (OFF), 3 = stuck ON and 4 = stuck OFF. These last two states are considered absorbing states, i.e. it is assumed that the components cannot be repaired once they fail. Moreover for each component the transition rates to states 3 and 4 are all equal, whatever may be the initial state 1 or 2. In this simple problem the fluid level is the only process variable and the eqs.(1) reduce to linear algebraic equations. Starting from time t_j^0 at which the system enters the configuration j and the fluid level is $H(t_j^0)$, at time $t \geq t_j^0$ we have

$$H(t_j^0 + t) = H(t_j^0) + (a_1 Q_1 + a_2 Q_2 - a_3 Q_3)(t - t_j^0) \tag{2}$$

where Q_i, for simplicity sake assumed to be constant, is the rate of level variation due to the i-th unit and

$$a_i = 1 \quad \text{if the unit is ON or stuck ON}$$
$$= 0 \quad \text{if the unit is OFF or stuck OFF}$$

In this simple case the set of vector eqs.(1) is represented by the scalar eqs.(2) for the evolution of the only physical variable $H(t)$: the vector j representing the system configuration is reflected on the a_i values as specified by the control laws of Table 1. Two process-variable-dependent top event conditions are considered: *dry out* and *overflow*. The mission time for the system is 1000 h and the objective of the analysis is the evaluation of the time-dependent probabilities of both *dry out* and *overflow* within that period of time. Figure 3 shows that the level may belong to three different regions. Region 2 is considered the region of "correct functioning" and the control operates in such a way to bring the level back within this region, as shown by the control laws of Table 1. Component 1 (e.g. a pump) provides the fluid to the container with a rate of level variation, $Q_1 = 0.6\ m/h$. Failures of this component are exponentially distributed with a time constant of 219 h. Component 2 is identical to component 1, except for its time constant being 175 h. Component 3 is an outlet element (e.g. a valve) whose flow rate also produces a level variation $Q_V = 0.6\ m/h$. The time constant for this component is 320 h. Notice that the fluid level takes approximately 3 hours to cover the entire region 2 so that the control is likely to be demanded upon a lot of times between successive stochastic transitions. At time $t = 0$ the system is in nominal configuration with components 1 and 3 operating and component 2 in stand-by; the fluid is inside the region of "correct functioning" at level $H(0) = 0$. In this situation the net flow rate is equal to zero and the fluid level remains constant until a stochastic transition occurs. The transition might modify the system configuration in such a way to make the level rise or lower. Each time the level enters a new region, the control system demands certain deterministic transitions (Table 1) to all the unfailed components in order to bring the level back in the region of "correct functioning" or at least keep it from going in *dry out* or *overflow*. If the control system works properly, the level remains in the new region or oscillates within region 2, until the next stochastic transition. The fluid level dynamic evolution is taken into account by comparing the time at which the fluid would enter a new region and that of the next stochastic transition. If the crossing to a new region should occur first, then unfailed units are demanded, by the control system, to appropriately change state. Otherwise the sampling of the stochastic event is carried out and the dynamic evolution is followed according to the new hardware configuration.

Table 1 Control laws for the system of Figure 3

Region	Level H	Unit 1	Unit 2	Unit 3
1	$H < HLA$	ON	ON	OFF
3	$H > HLB$	OFF	OFF	ON

4 Results

In all the following Monte Carlo simulations the number of trials is 10^5, unless explicitly specified.

The unavailability characteristics of the tank, related to *dry out* and *overflow*, have been evaluated both by a classical fault tree analysis and by a dynamic Monte Carlo approach. The classical fault tree analysis is based on the identification of the minimal cut sets for the two top events and their quantification in terms of probabilities. Table 2 reports the hardware configurations of minimal cut set leading to dry out and overflow. The computation has been done analitically and then checked via a "static" analog Monte Carlo simulation.

Table 2 Minimal cut sets for the fault tree methodology applied on the holdup tank of Figure 3

Component	1	2	3	Top event
mcs(1)	S.ON	S.ON	–	Of
mcs(2)	S.ON	–	S.OFF	Of
mcs(3)	–	S.ON	S.OFF	Of
mcs(4)	S.OFF	S.OFF	S.ON	Do

- mcs(i) = i-th minimal cut set
- S.ON(OFF) = Stuck-ON(OFF)
- Of = Overflow
- Do = Dry out

Figure 4 shows the results of the static and dynamic approach. As expected, the two situations are indistinguishable: indeed the effect of the control is to switch OFF and ON the components but this switching does not affect the failure occurrence since the transition rates have been chosen equal for the two states. This is a typical case in which the process variables variation and the control do not play any role on the reliability characteristics of the system which can be satisfactorily analyzed through the classical methodologies, with spare of CPU time.

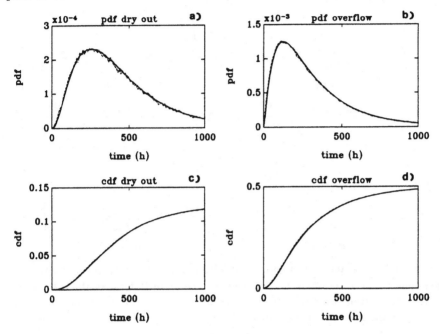

Figure 4 Holdup tank with no failures on demand and $Q_1 = Q_2$:
a) Dry out probability density function;
b) Overflow probability density function;
c) Dry out cumulative distribution function;
d) Overflow cumulative distribution function;
Solid line represents the analytical solution, dashed line the static analog Monte Carlo and dotted line the dynamic analog Monte Carlo.

The situation is quite different when the failure rates of the components in state ON differ from those in state OFF: then the above mentioned switching due to the control does affect the reliability of the system. This is shown in Figure 5 which relates to the case in which the failure rates of the components in states opposite to the initial ones have been increased by a factor of 10 for the transitions towards state #3 (stuck-on) and 100 for transitions towards state #4 (stuck-off). The dynamic approach is also needed when failures in the control are considered. Let us consider the tank in which there is now a certain probability of failure of the components to correctly respond to a demand by the control. We assume that this kind of failure leaves the component in its state ON or OFF, as if the demand were never done.

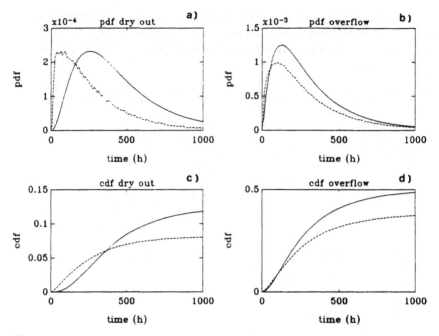

Figure 5 Same example of Figure 4 but with failure rates of the components, in state opposite to the initial ones, increased by a factor of 10 for transitions to state #3 and 100 for transitions to state #4. Solid line represents the analytical solution, dashed line the dynamic analog Monte Carlo.

From a different point of view, we could think of a human–operated control system in which an operator switches the states of the components according to the control laws. A failure then occurs if the operator "forgets" to switch the component in the correct state as required by the situation. These kinds of errors might be included in the category of "slips", i.e. errors of omission, as described by [12], [20]. In our simple model we assume to have the operator that no one would like to have in a nuclear plant. This operator has a 10 % probability of forgetting to make the proper actions to respond to each demand. In this situation, the evolution of the process variables (in our case the tank level) and the consequent actions of the control have a very strong influence on the unavailability of the system. Indeed, Figure 6 shows that both top events, dry out and overflow, occur earlier in time and with increased cumulative probabilities, compared to the case with no failures on the control. The presence of these differences even in such a simple case is an argument in favour of a dynamic reliability analysis even if the required efforts (software complexity and CPU time) are undoubtedly larger. The dynamic approach has also been compared with the well–known DYLAM methodology [21]. The results show a good agreement between the two with reference to the tank problem of Figure 4: in this case component #2 has a flow rate $Q_2 = Q_1/2$.

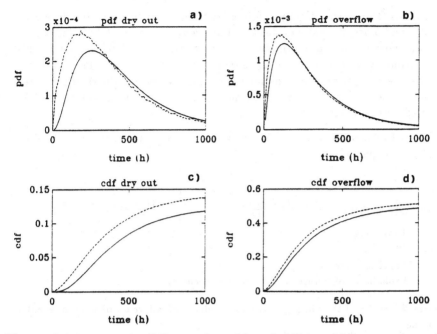

Figure 6 Same example of Figure 4 but with probabilities of failures on demand equal to 0.1 and the range of correct functioning equal to $\frac{1}{100}$ that of the normal case. Solid line represents the static analytical solution while dashed line represents the dynamic analog Monte Carlo.

5 Forcing the Failures on Demand

When treating higly reliable systems, the analog Monte Carlo method has the major drawback that most of the histories end without giving information on the failures. In order to face this problem, biasing techniques have been introduced [7,8],[22–24]. However one should be aware that favouring certain configurations unavoidably corresponds to unfavouring others. When these latter also contribute to the desired quantity, their large weights may deteriorate the resulting statistics. In the dynamic approach this consideration may be of importance when biasing the control failures since the control's effect is entangled and therefore hard to guess. Moreover if one forces in a particular direction but he is also interested in the consequences of the unfavoured events, then the results will generally have a poor statistics.

We can better explain the above point by considering a very simple system consisting of a tank with just one inlet component, for example a pump. The fluid has an initial random level between 0 and H. At time $t = 0$ the pump is turned ON and the level consequently rises. When the level fills the container up to the top (H) then the control demands that the pump be turned OFF. If the control operates successfully, event 2 is said to have occurred; if it fails, then event 3 (which is an overflow) is reached. Event number 1 occurs if the pump fails OFF stochastically before filling the container up to H. Note that event 2 and 3 are mutually exclusive while event 1 is completely independent of both. Forcing the failure on demand due to the control shows how in this case favouring an event may lead to worse estimates of the others. Table 3 reports the analytical equations for the cumulative probabilities of the three events, while Table 4 reports some results obtained by forcing the failures on demand, with 10^4 trials.

Table 3 Probabilities of the three top events for the simple example of the tank with just one pump

Event i	1	2	3
$prob.(i)$	$1 - \frac{v}{\lambda H}(1 - e^{-\frac{\lambda H}{v}})$	$1 - \frac{p_2 v}{\lambda H}(1 - e^{-\frac{\lambda H}{v}})$	$1 - \frac{(1-p_2)v}{\lambda H}(1 - e^{-\frac{\lambda H}{v}})$

- H = Tank Height
- v = Filling rate
- λ = Transition rate
- p_2 = Probability of failure on demand ($\rightarrow$ event 2)

Table 4 Results of the biasing of the probabilities of failure on demand of event 2 of Table 3

Event	1	2	3
Analytic	$1.239634 \cdot 10^{-2}$	$2.962811 \cdot 10^{-1}$	$6.913225 \cdot 10^{-1}$
Analogic	$1.240000 \cdot 10^{-2}$	$3.019000 \cdot 10^{-1}$	$6.857000 \cdot 10^{-1}$
Bias $p_2 = .3 \rightarrow .9$	$1.240000 \cdot 10^{-2}$	$2.957544 \cdot 10^{-1}$	$7.021000 \cdot 10^{-1}$
Bias $p_2 = .3 \rightarrow .999$	$1.240000 \cdot 10^{-2}$	$2.963835 \cdot 10^{-1}$	$4.200054 \cdot 10^{-1}$
Bias $p_2 = .3 \rightarrow 1$	$1.240000 \cdot 10^{-2}$	$2.963835 \cdot 10^{-1}$	$0.000000 \cdot 10^{+0}$

- Tank Height, $H = 10$m
- Filling rate $v = 2$m/h
- Transition rate $\lambda = 0.005h^{-1}$
- Natural probability of failure on demand (event 2) $p_2 = 0.3$

Concerning the holdup tank of Figure 3, if one forces the control failures in order to get a better estimate of the top event probabilities, this actually may lead to unsatisfactory results, as shown in Figures 8, 9 and 10 in which the probabilities of failure on demand have been forced from 0.1 to 0.3, 0.9 and 1.0 respectively. It appears that a biasing of 0.3 essentially worsens the statistics (Figure 7); that of 1.0 changes the gross shape of the results due to the fact that some control actions are prohibited and therefore their contribution is missing (Figure 9); the intermediate case of 0.9 suffers of a worsened statistics which also produces a shape distortion (Figure 8). This is confirmed by the results of Figure 10 representing the same case of Figure 8 with statistics improved by a factor of 100 and consequent reduction of the shape distortion.

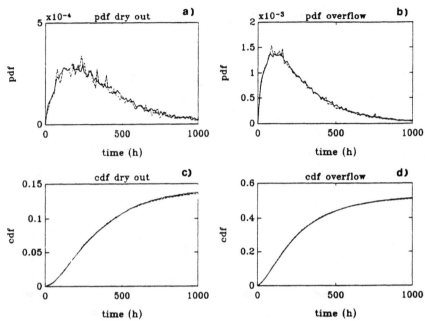

Figure 7 Same problem as in Figure 6 but with the probabilities of failures on demand forced from 0.1 to 0.3. Solid line represents the case of Figure 6 (0.1) and dashed line represents the biased case (0.3).

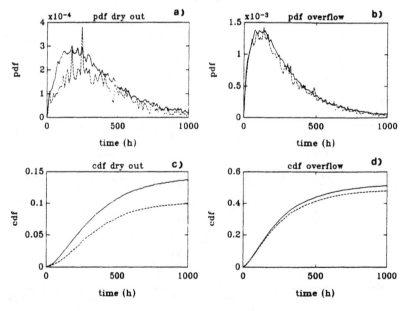

Figure 8 Same problem as in Figure 6 but with the probabilities of failures on demand forced from 0.1 to 0.9. Solid line represents the case of Figure 6 (0.1) and dashed line represents the biased case (0.9).

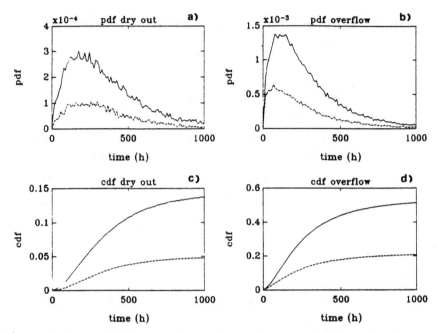

Figure 9 Same problem as in Figure 6 but with the probabilities of failures on demand forced from 0.1 to 1.0. Solid line represents the case of Figure 6 (0.1) and dashed line represents the biased case (1.0).

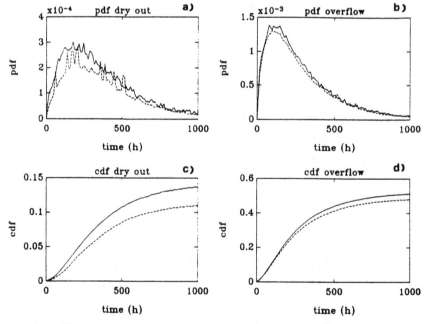

Figure 10 Same problem as in Figure 8 with 10^7 trials.

6 Conclusions

In this paper we presented a way in which the Monte Carlo approach to reliability analysis might be extended to treat dynamic systems. The approach has been seen to have the potentiality to include many of the issues addressed by a realistic analysis. The simple literature case of the holdup tank has been discussed. The results show that in some cases the classical fault tree/event tree analysis and the dynamic approach do not differ. This suggests that in general a preliminary investigation on the actual need for a dynamic PRA is in order. However the results have also shown that the process variables and the protection/control system may actually influence the analysis. A comparison with the DYLAM methodology has shown good agreement in the results. Failures on demand, or operator's slips, have been shown to increase the influence of the process variables and control system on the unavailability characteristics of the plant. In these cases, performing a dynamic analysis leads to more realistic results.

The experience gained from simple cases such as the one here presented indicates that in real plants the complexity of the systems may lead to extremely large computer times required by the dynamic analysis and therefore many research efforts are still needed to develop quicker physical models and more efficient techniques for their evaluation. Specifically in the Monte Carlo simulation, biasing techniques are most welcomed but they should be implemented with particular care. In any case, the computer used and the CPU time required should be given in order to allow a proper comparison among different methodologies. Moreover the propagation of the uncertainties of the initial data should be carefully evaluated: this is particularly true whenever an operator's model is introduced, because of the large uncertainties characterizing any human behaviour, especially under stress conditions.

References

1. L. Olivi, Response Surface Methodology. Handbook for Nuclear Reactor Safety, EUR 9600, EN, 198
2. C. S. Hsu, Cell-to-Cell Mapping: A Method of Global Analysis for Non-Linear Systems, Springer-Verlag, New York 1987.
3. M. Belhadj, M. Hassan, T. Aldemir, The Sensitivity of Process Control System Interval Reliability to Process Dynamics– A Case Study, PSAM Proceedings, Elsevier Publishing Company Co., Inc. 1991, pp.533–538.
4. T. Aldemir, Utilization of the Cell-to-Cell Mapping Technique to Construct Markov Failure Models for Process Control Systems, PSAM Proceedings, Elsevier Publishing Company Co., Inc. 1991, pp.1431–1436.
5. M. Marseguerra, S. Minoggio, A. Rossi, E. Zio, Neural Networks Prediction and Fault Diagnosis Applied to Stationary and Non-Stationary ARMA Modeled Time Series, submitted for publication in Progress in Nuclear Energy.

6. S. J. Kamat, M. W. Riley, Determination of Reliability Using Event–Based Monte Carlo Simulation, IEEE Trans. on Reliability, Vol. R–24, Apr. 1975, pp.73–75.

7. H. Kumamoto, K. Tanaka, K. Inoue, Efficient Evaluation of System Reliability by Monte Carlo Method, IEEE Trans. on Reliability, Vol. R–26, N.5, 1977, pp.311–315

8. E. E. Lewis, F. Bohm, Monte Carlo Simulation of Markov Unreliability Models, Nucl. Eng. and Des. 77, 1984, pp.49–62.

9. E. E. Lewis, T. Zhuguo, Monte Carlo Reliability Modeling by Inhomogeneous Markov Processes, Reliability Engineering & System Safety, Vol. 16, 1986, pp. 277–296.

10. T. Zhuguo, E. E. Lewis, Component Dependency Models in Markov Monte Carlo Simulation, Reliab. Eng. & System Safety, Vol. 13, pp.45–61, 1985.

11. J. Devooght, C. Smidts, Probabilistic Reactor Dynamics I. The Theory of Continuous Event Trees, Nucl. Sci. and Eng. 111, 3 (1992),229–240

12. N. Siu, Risk Assessment for Dynamic Systems: An Overview, submitted for publication to Reliab. Eng. & System Safety, 1992.

13. W. E. Vesely, A Time Dependent Methodology for Fault Tree Evaluation, Nucl. Eng. and Des. 13, pp.337, 1970.

14. T. Aldemir, Computer–Assisted Markov Failure Modeling of Process Control Systems, IEEE Trans. on Reliability, Vol. R–36, N. 1, pp.133–144, 1987

15. D. L. Deoss, N. Siu, A Simulation Model for Dynamic System Availability Analysis, MITNE–287, Mass. Inst. of Tech., October 1989.

16. A. S. Householder, G. E. Fosythe, H. H. Germond Eds., Monte Carlo Method, NBS, Applied Mathematics Series, 12, 6, 1951

17. G. Goertzel, M. H. Kalos, Monte Carlo Methods in Transport Problems, Progr. Nucl. Energy Ser. I, 2, pp.315–369, 1958.

18. M. H. Kalos, P. A. Whitlock, Monte Carlo Methods Vol. I: Basics, John Wiley & Sons, 1986.

19. V. N. Dang, D. L. Deoss, N. Siu, Event Simulation for Availability Analysis of Dynamic Systems, SMiRT 11, Tokyo, Aug. 18–23, 1991.

20. P. C. Cacciabue, G. Mancini, U. Bersini, A Model of Operator Behaviour for Man–Machine System Simulation, Automatica, 26, pp.1025–1034, 1990.

21. P. C. Cacciabue, A. Amendola, G. Cojazzi, Dynamic Logical Analytical Methodology Versus Fault Tree: The Case Study for the Auxiliary Feedwater System of a Nuclear Power Plant, Nucl. Tech., 74, pp. 195–208, 1986.

22. H. Kumamoto, K. Tanaka, K. Inoue, E. J. Henley, Dagger–Sampling Monte Carlo for System Unavailability Evaluation, IEEE Trans. on Reliability, Vol. R–29, N.2, pp.122–125, 1980.

23. M. Marseguerra, E. Zio, Non Linear Monte Carlo Reliability Analysis with Biasing Towards Top Event, accepted for publication in Reliab. Eng. & System Safety, 1992.

24. C. Smidts, J. Devooght, Probabilistic Reactor Dynamics II. A Monte Carlo Study of a Fast Reactor Transient, Nucl. Sci. and Eng. 111, 3 (1992),241–256.

Dependability Analysis of Embedded Software Systems

C. T. Muthukumar, Sergio B. Guarro and George E. Apostolakis

Mechanical, Aerospace and Nuclear Engineering Department
University of California, Los Angeles, CA 90024-1597, U.S.A

Abstract. A two-step Dynamic Flowgraph Methodology (DFM) is presented for the dependability analysis of embedded systems. The first step in its application consists of building a model that expresses the logic and dynamic behavior of the system in terms of its physical and software variables. The modeling framework that is used combines and expands on the structures of two known techniques: logic flowgraph methodology (LFM) and Petri nets, the former being used to model flow of causality and the latter to model the timed control of the causality flow. The second step of DFM application consists in using the model developed in the first step to build "timed" fault trees that identify and represent logic combinations and time sequences of variable states that can cause the system to be in certain specific states of interest (desirable or undesirable). This is accomplished by backtracking through the DFM model of the system of interest in a systematic, specified manner. The information, contained in the fault trees, concerning the hardware and software conditions that can lead to system states of interest can be used to uncover undesirable or unanticipated software/hardware interactions and to improve upon the system design by eliminating unsafe software-execution paths. It can also be used to develop a focused testing strategy.

Keywords. Embedded systems, Logic Flowgraph Methodology, fault trees, Petri nets, software dependability.

1 Introduction

Embedded systems can be defined as those in which the functions of mechanical and physical devices are controlled and managed by dedicated digital processors. The increased use of computers in modern aircraft, nuclear power plants, chemical plants, missiles, and space systems to control most system functions has brought forward serious issues of software (in addition to hardware) dependability. A system is dependable when it properly addresses the issues of safety, reliability, security, etc. [1]. The flexibility provided by the software implementation of system control functions has extremely strong appeal to the system designer, but also carries a potentially dangerous backside. Unlike hardware (which is usually limited in complexity by its own functional and physical constraints), software can in fact be made as complex as the designer wishes, with potentially daunting issues of dependability and quality assurance ensuing if this complexity potential is not properly controlled and dealt with. In embedded systems, software logic errors in design or programming may cripple the operation of the whole embedded system. Events of this nature have indeed occurred causing severe consequences such as loss of human lives or large economic losses. Increased system complexity and severe loss potential has thus created a strong motivation for developing tools and techniques by which embedded systems can be analyzed from the point of view of determining and ensuring their reliability and safety attributes.

According to whether the objective is software reliability or software safety, the analytical approaches developed and followed differ considerably in practical terms. Software reliability has come to indicate the formulation of statistical models for predicting error frequencies [2], while software safety focuses on ways to uncover severe-consequence errors [3-5]. Here, we are concerned with analyzing embedded systems from the perspective of ensuring the dependability of their key functions. Our approach is to identify critical paths that lead to certain specific system states of interest and that can be applied generally to either normal or faulted system states.

Traditional reliability theory has been successful in modeling and analyzing the impact of random or "wear-out" failures of physical components on overall system reliability. It has, however, been less successful in modeling human and design errors, which characterize the nature of essentially all of the faults that occur in software. An important difference between hardware and software failures is that, unlike hardware components, a software system does not age or deteriorate [6] (although sometimes software may appear to age as certain faults are exposed in the software by the aging and deterioration of connected hardware). The typical way in which software affects the system dependability is that "dormant faults" in the software become actual failures when the normal execution flow or the occurrence of an unanticipated set of external conditions causes them to manifest themselves, thus affecting the overall system behavior. Because software errors of an intrinsically trivial nature (such as the change of a single byte of data from its true value) may result in drastic upsets in the observable system behavior, it becomes all the more important to ensure its dependability in relation to its functions in the embedded system.

Software is one of the most complex creations of human kind and software errors occur due to our inability to comprehend the intricacies of these complex products [7]. Several methods, which include software fault-tree analysis [8,9], static and dynamic testing [10], Petri net analysis [11], and mathematical verification [12,13] have been proposed for analyzing software for its reliability and safety features. Extensive random testing, although capable of uncovering many errors in software, cannot guarantee that all critical errors are uncovered [10] consistently with the principle of functional testing (i.e., guided by the identification and coverage of all software functions) [10], our methodology proposes a complement to testing that identifies critical parts of the software upon which to focus attention.

The methods used to analyze hardware in complex systems include fault tree analysis, digraphs [14], decision tables [15-17], logic flowgraph methodology [18-21], dynamic logical analytical methodology (DYLAM) [22-24], equation "bigraphs" [25], and functional equations [26]. The end result of almost all of these methods (except for DYLAM) is fault trees that show how a certain identified state of the system can occur. DYLAM simulates system evolution in time by solving differential equations (using the finite difference method) describing the system.

In all critical applications (and especially in those which are safety-critical), it is extremely important that the system software should behave as expected for all possible demands and input conditions created for it by the interfacing hardware and physical world [2]. Hence, the execution of an embedded-system safety-and-reliability analysis should be guided by the idea that, in these systems, software and hardware cannot be viewed in isolation, but only from the vantage point of a well-integrated understanding of the overall system functions and interfaces. It is also important, in the analysis of embedded systems, to realize that potentially-critical timing requirements need to be accounted for and that software computations upon which the system control must rely are based on the history of measured data. It is thus imperative that the time element be explicitly included in any embedded system analysis. The above considerations have formed the basis and are the guiding principles in the development of our methodology.

2 The Model Building Process

The methodology proposed follows a two-step process:

Step 1: Building a model of the embedded system that expresses the logic of the system in terms of the causal relationships among hardware and software variables, as well as the temporal characteristics of the execution of software modules.

Step 2: Determining how the embedded system can attain a certain state (desirable or undesirable), i.e., the combinations and sequences of physical and software variable states that can cause the system to attain that top-level state.

The model building process is illustrated by using the simple embedded system shown in Fig. 1. The example is one of a hydraulic flow-rate control-valve whose position is controlled by the feedback control software to maintain a constant flowrate (3.50 gal/min) in the pipe. The pipe is fed from a storage tank filled with water to a height H (initially 5 m). The feedback control software runs in the digital control module (DCM) and takes in, at constant time interval time increase, the measured value of the downstream flowrate every Δt and then, according to its programmed logic, gives output signals that control the valve position and, hence, the flow rate itself. DCM software, shown in Table 1, compares the measured flow rate with a reference set point (the desired constant flow rate) to calculate the corresponding the error and the fractional error. Then, the new valve position control (VC) signal is calculated based on the valve position in the previous cycle (VXP) and the fractional error. The programmed logic is a proportional-integral control logic, and the code corresponding to it is shown in Table 1. Given that the reference flowrate is 3.50 gal/min, if the measured flow rate is less than 5.50 gal/min or greater than 1.75 gal/min, then the valve position is not changed. If the measured flow rate greater that 5.50 gal/min, then the valve is closed down by one position. If the measured flow rate is less than 1.75 gal/min then the valve is opened by one position. Checks have also been built in to ensure that control signals cause the new valve position signal to be kept within physical limits.

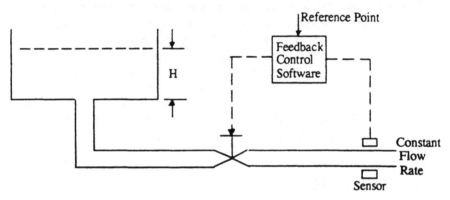

Fig. 1 Simple Embedded System

An important point of the DFM methodology worth noting is that modeling constructs for the hardware and software are similar, although some differences exist in the processing steps to transform the information available about them into the structure of modeling constructs (the software part requiring a few additional intermediate steps). The final model is obtained by interfacing the hardware and software parts of the model. The remainder of this section describes in detail the steps to be executed in the construction of both portions of a DFM model and illustrates these steps for the flowrate control process just described above.

4.1 Hardware Model Building

Step 1: Identification of process variables.

Identify the necessary and sufficient variables that are required to describe the system accurately and effectively. These identified variables will form the nodes in the model. In many cases, the relationships between physical variables of the system can be different for different states of a hardware component. Hence, nodes representing the state of a component, which affect relationships between physical variables which are called conditioning nodes, are also identified and represented in the model. These conditioning nodes represent healthy vs. degraded states of hardware components that affect system behavior. Each DFM node, regular or conditioning, represents a vector of the discrete states that are chosen to represent the status of the corresponding system parameter or variable at any given time.

The model (Fig.2) is built with physical variables H, F, VX, and VXP as nodes. The modeling equation used is

$$F = Constant*VX* \sqrt{D*g*H}$$

where

H = Height of water column

D = Density of water

VX = Valve position

g = Acceleration due to gravity

Table 1   Digital Control Module (DCM) - Feedback Control Software

REF=3.50

VXP=VC

ER=FM-REF

ERF=ER/REF

IF(ERF>1.86)THEN ΔVC=-0.5

IF(0.49<ERF≤1.86)THEN ΔVC=-0.25

IF(0.0≤<ERF≤0.49)THEN ΔVC=0.0

IF(-0.43≤ERF<0.0)THEN ΔVC=0.0

IF(-1.00≤ERF≤-0.43)THEN ΔVC=0.25

IF(ERF<-1.00)THEN ΔVC=0.5

VC=MIN(VXP+ΔVC,1.0)

VC=MAX(VXP+ΔVC,0.0)

REF = Constant flow rate to be maintained in the pipe

FM = Flow rate measured

ER = Difference between the measured flow rate and the refernce point

ERF = Fractional error with respect to the reference point

VC = Valve control

VXP = Valve position in the previous cycle

Step 2: Identification of causality flow

Develop a network that shows the causality flow (which variables are affected by which other variables) and any time dependencies among the hardware nodes. The network takes the form of directed graphs, with relations of causality and conditioning switching action represented by directed edges that connect the network nodes. The causality edges connect the network nodes through transfer boxes or transition boxes. The conditional edges input, into the appropriate transfer boxes, the states of hardware components or other system "logic switches" that may affect the relationship between the output and the input variables (conditioning nodes). Therefore, conditional edges connect the conditioning nodes to the variables that they affect through the appropriate transfer or transition box. The relationships between the input and output node(s) are represented in by decision tables in a transfer or transition box. In the case of transfer boxes, the association between the input and output states is assumed to exist in the same time frame, i.e., there exists an "instantaneous correspondence." In the case of transition boxes, there is a lag between the time when a variable state(s) becomes true and the time when the output variable associated with these inputs is reached.

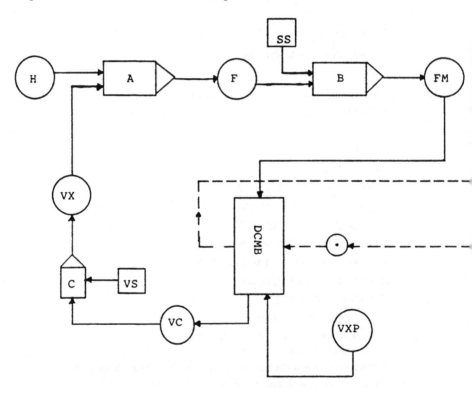

H = Height of water column
F = Flow rate
SS = Sensor state
FM = Flow measurement
VX = Valve position at the present time

VS = Valve state
VC = Valve control
A, B, C = Transfer boxes
DCMB = Digital control modeling box
VXP = Valve position in the previous cycle

Fig. 2 Model of the Embedded System

In the embedded system example, *FM* is affected by node *F* and conditioning node *SS*. Hence, this relationship is represented by transfer box *B*, which takes *F* and *SS* as inputs and gives out *FM* (Fig.2). This transfer box *B* contains only one decision table, since it has only one output. Similarly, other relationships between the physical variables are represented in the model and, hence, the causality network is developed.

Step 3: Definition of process variable state vectors.

Define a vector of finite number of discrete values that characterize the possible range that each node is expected to take. Discretize the range of physical variables, keeping in mind the physical significance of the variable and the minimum number of states needed to characterize its status. Discretizing it into too many states increases the size of the downstream transfer/transition box decision tables considerably (see also discussion in Step 4 below).

In the case of the conditioning nodes, since they represent the different states of a component or a specific logic condition in the process, the discretization is more direct and defined. For example, for a physical component it is done by enumerating the number of possible states of the component. Thus the possible states of conditioning nodes SS and VS are:

SS [Fails high(1), Normal(0), Fails low(-1)],

VS [Fails open(1), Normal(0), Fails closed(-1)].

Step 4: Definition of transfer/transition box decision tables

Develop decision tables for each output node of every transfer box and transition box. Each decision table is a simple mapping between the combinatorial states of inputs (including conditioning inputs) to the output. The first step in developing decision tables is to identify the inputs (including conditioning inputs) that affect the chosen particular node (output) for which the decision table is being developed. This is done by inspecting the model network that explicitly shows the causality relations among the different variables. For each input, a vector of finite discrete values (states) has been defined (per Step 3) that characterizes the possible range that the input variable is expected to take. Each of these discrete values is representative of a range of continuous values around it. One can then enumerate all combinations of input states and the corresponding output and order the corresponding state combinations in a two-dimensional table. Table 2 shows a decision table for the model shown in Fig. 2. For example, in Table 2 the inputs are *VX* and *H*, and the output is *F*. The table has 3 columns, one for the output and one for each input. Since *H* is discretized into 4 values and *VX* into 5 values, the number of rows in the table is 4x5= 20. The first row in the table states that when *VX* =0 and *H* =0.0 the output value is *F* =0.0. Similarly, the last row states that when *VX* =1.0 and *H* =5, the output is *F* =10.0.

Table 2. Decision table in Transfer Box A

Inputs		Output
VX	H	F
0.0	-	0.00
0.25	1	1.00
0.25	3	1.50
0.50	1	2.00
0.25	5	2.50
0.75	1	3.00
0.50	3	3.50
1.00	1	4.00
0.50	5	4.75
0.75	3	5.25
1.00	3	6.75
0.75	5	7.50
1.00	5	10.00

Step 5: Identification of transfer/transition box fault conditioning

In the hardware part of the system, physical relationships between nodes when a system component is in a faulted state can be different from when it is functioning normally. Also, relationships between nodes can change according to the system mode of operation ("system logic switches"). This information is included in the DFM model through the use of conditioning nodes which represent either the normal/faulted states of a system component or other "system logic switches". According to the state of a conditioning node, the relationship between nodes that it affects may be different and is represented by a separate decision table for each state of the conditioning node. The network which shows how conditioning nodes affect relationships between nodes is called fault conditioning. It is important to include fault conditioning in the model, because it reflects changes in the different possible relationships between nodes and hence can be used to analyze how it affects the overall system logic and behavior.

4.2 Software Model Building

The model-building steps for the software portions of the embedded system follow a rationale that is quite similar to that employed to model the hardware and physics. Particular attention, however, must be dedicated to the tasks of identifying key system variable representations in the software itself, and the time effects and dependencies existing in the execution of the software process.

Step 1: Identification of software time components

Identify the time aspects of the software. Embedded system software, generally, is real-time control software that receives real-time data from the physical part of the system and performs appropriate control actions on the basis of this data. The first step in DFM software model building is to identify the portions of the software that specifically involve time dependencies (e.g., interrupts or synchronization of routines), and develop a Petri-net kind of structure to represent these time elements in the model.

Identification of these time elements is made by partitioning the entire software into execution blocks called software time components (STC) to expose all time dependencies associated with the software execution. In most cases, most time dependencies are controlled in the main controlling module, however, modules other than the main module can also include time and synchronization elements. In such cases, such additional time elements should be isolated as STCs. Software execution flow-diagrams should be used by the analyst while proceeding to the identification of STCs. In the DFM models, all STCs will be represented by transition boxes and an associated Petri-net like structure reflecting timing or synchronization logic and the time sequence characteristics of execution (see Step 3).

In the example shown in Fig. 1, the principal timing mechanism consists of a cyclically repeated interrupt every Δt, which causes the DCM to execute. DCM is thus an STC and is modeled using a Petri-net like construct. This, in turn provides a reachability graph which shows the relative order of software execution, in our case simply expressing the cyclical updating of software variable VC, based on the value of the variable FM at time $t = t_n$ and on the value of the variable VXP, which is equal to the value of the variable VX at time $t = t_{n-1}$.

Step 2: Identification of software functional components and key software variables

The next step is to break up the STCs (including the parts of the code not involving time dependencies and with no time element involved other than their own execution time) into smaller parts, so that each part represents a physically meaningful action. Each of these parts of the code can be considered a "software functional component" (SFC). The analyst should use the specifications, data dictionary, and the structure chart to decide on the definition of software components. Each identified SFC forms a transfer box in the DFM model. The general guideline would be to check whether "procedures" of the main program can form software components without further breaking. This is because, in modern software engineering practice, designers try to encapsulate physically meaningful actions into identifiable, separate modules. This general guideline, of course, does not mean that one should never transgress the boundaries of modules in deciding what constitutes a software functional component. If a module performs several functions as a "super component", it would be practical to look at it and further subdivide it into smaller SFCs. An example would be a module that is switching actions according to its tag value. In this case, the part of software corresponding to each tag value can be considered a sub-component of the module and forms an SFC. In order to reduce the complexity of the model without compromising accuracy, standard library routines need not be considered as separate software components by themselves but rather as a part of the software component that actually calls them.

Step 3: Identification of software component time dependencies and control flow

Petri-net-like structures are used to model the time elements and the control flow among the various software components. In the example shown, we have one software component, DCMB, and the timing mechanism that executes DCMB every Δt after sampling FM. The DCMB is represented in the model as a transition box, which gives an output token Δt after it is enabled (Fig. 2).

Step 4: Identification of software component data flow

In addition to control flow, the data flow between software components must be shown in order to give a complete representation of the software system, in addition to control flow. This step is analogous to Step 2 in the hardware model-building, which identifies "causality flow" by showing which node is input/output of which transition or transfer box. Data flow representation is required because, in the analysis of the model that is performed to build fault trees, the information about the failure modes in software components is naturally expressed as the occurrence of "faulted states" in certain data elements. The input and output data elements of software components constitute data flow, and causality flow, in the software portion of the model. Data flow is represented as a network of directed edges connecting software nodes through transfer or transition boxes (software components). In our example, the software node, *FM*, is linked as input into the transition box, DCMB, and the software node, *VC*, is linked as output from it, where *FM* is the measured flow rate and *VC* is the valve control.

Step 5: Definition of key software variable state vectors

For each of the key software variables identified, it is necessary to define a vector of discrete values that characterizes the possible range that the parameter is expected to take. This is initially done for the software images of the physical parameters input into the software from the hardware. Then, these values are propagated through the code to find the corresponding discretized ranges for other variables. In our example, the input FM to the software will be characterized by discretized values between a minimum value of zero and a maximum possible value for that system (FM=10.0). Simple error checks can be placed to detect inputs outside this range.

Step 6: Definition of software component decision tables

A decision table is built for each output from a software component (transfer or transition box). Once the discretized ranges of the input parameters to a software component have been identified, a decision table is built that is a mapping between the different combinatorial states of input and its corresponding output value. The output value for a particular combination of inputs is found by executing the piece of code corresponding to the software that has that particular combination of inputs as the initial values of the input parameters.

Step 7: Definition of software component fault or mode conditioning

In software, situations in which different software logic paths are taken depending on, say, an index or tag value, or whether a certain condition is satisfied are commonplace. Taking a different software path or another can result in drastic differences in the overall system behavior. Hence, in order to build a model that would represent the system behavior with high fidelity, it is extremely important to represent conditions that cause different paths to be taken in the software. This is done using a fault or mode conditioning network. In the case of software, mode conditioning would be generally expected to be more common than fault conditioning.

In many cases, the fault or mode conditioning may be expressed in terms of low-level internal variables, which may not be represented in the data flow of the DFM model. Since the state of these low-level internal variables drastically influences the system behavior, it becomes important to backtrack to find their root causes. This can be done by performing a standard Software Fault Tree Analysis (SFTA) [8,9]. The backtracking is continued until software variables which are images of physical parameters are reached back into the DFM network.

4.2 Hardware/Software Model Integration

The principal step in integrating the software and hardware portions of the embedded system model is the identification of all data that are input and output from the software system to the hardware. Then, corresponding parameters in the software and hardware model are connected, with the direction of flow properly indicated. In almost all cases, an interface point is constituted by either a sensor that measures the physical parameter before its value is input into the software has to be included in the model, or by an actuator that takes a software-derived value for the desired adjustment to be made in the position/state of a hardware component and makes the hardware respond to it. Therefore, in the case of a sensor, a node is to be shown in the DFM model for the measured physical parameter and one for the measurement out of the sensor, or, in the case of an actuator, a node must be used for the control signal input to the actuator, and another node for the actual hardware parameter which characterizes the response of the actuator has to be included in the hardware portion of the model. The sensor or actuator themselves are modeled as transfer boxes or transition boxes, as appropriate, and with conditioning inputs if the sensor or actuator may have any degraded states. In Fig.2, the software and hardware parts of the model are interfaced by connecting VC and VX through transfer box C with conditioning input VS, and by connecting F and FM through transfer box B with conditioning input SS.

3 Timed Fault-Tree Construction

The product of embedded system analysis are the fault trees that are derived from the model. Hence, the fault-tree construction algorithm is the backbone of the methodology. The flowchart of this algorithm has been divided into five steps to better explain the way it works.

Step 1:

Identify the top event (desirable or undesirable state of the system) of interest and translate it into the state(s) of one or more nodes (variables) of the system. Each identified variable and its corresponding state is referred to as a "var_to_be_backtracked" (event). If there is more than one var_to_be_backtracked, they are placed under an AND gate and are developed individually. Backtracking through the model begins with the node corresponding to var_to_be_backtracked.

Step 2:

Derive the reachability graph of the transition boxes from the DFM model. In the case of real-time systems, the behavior of the system depends on the order in which transition boxes are fired [27], hence they must be determined. This information is necessary in order to develop the fault tree. In general, the terminal event of each branch of the fault tree, which is developed completely for a discrete point in time, will be the output of a transition box or a basic event (an event that cannot be developed further). In order to determine which terminal event(s) to develop next, one has to know which transition box fired last. This information is provided by the reachability graph.

Step 3:

Start backtracking. This can be done according to procedures similar to those used in LFM [18]. If at any point in the backtracking process, an output of a transition box or a basic event is reached, development of that branch of the fault tree is suspended and backtracking continues in the current time frame for the remaining branches that have not been developed until a basic event or an output of a transition box is reached for these as well. After all the branches have been developed, backtracking continues through the most recent transition box. At a given point in time, the events corresponding to the output nodes of the most recent transition box are located and are developed completely before backtracking is allowed through next transition box (i.e. moving to the previous time frame), as explained in Step 4. The result is a series of fault trees that represent the state of the system at different times and are linked through the inputs and outputs of transition boxes. The logic utilized in Step 3 for the construction of static fault trees from the decision tables closely resembles that used in the LFM [18,19] and

Special note on Consistency Checking

This part of the logic uses consistency checking to ensure that events newly placed in the tree are consistent with other events already in the tree. Branches containing inconsistent events are pruned from the tree. Consistency checking is one of the most important steps in developing the fault tree since it ensures that impossible events are pruned immediately, which saves the effort of developing these impossible events and reduces the size of the tree exponentially. In fact, consistency checking makes the fault tree a useful analysis tool by ensuring that only meaningful information is ordered in the tree. The consistency checking logic, as implemented in LFM [18,19] and CAT [15,16], only checks against static relationships between the variables. However, since the fault trees for embedded systems will contain information about the system at different points in time, it is necessary also to perform consistency checks against dynamic relationships between the variables.

Step 4:

Once, the fault tree is developed completely for a given discrete point in time, in order to develop the tree further, the transition that fired prior to the present discrete point in time should be known. This information is provided the reachability graph obtained via the Petri net analysis. This step is important since the behavior of the system depends on the order in which transition boxes are fired.

Step 5:

Once the static fault tree for the current time has been fully developed and the last transition box to have fired has been identified, the backtracking procedure continues by identifying the input states that could cause the output of the last transition box. The input-output relationships for the transition boxes are modeled by decision tables, just as they are for transfer boxes. Based on the output states of the last transition box, the input states are obtained from the decision tables. Then, static sub-trees are developed for the identified input states to the last transition box. The procedure then continues following Steps 3-5.

The procedure just defined above is illustrated in the following, relative to the analysis of a specific event of interest for the simple embedded system depicted in Fig. 1, using the DFM model of this system shown in Fig. 2. The top event of interest is "Flow rate is too high" and the timed fault tree obtained by applying the backtracking procedures described above is shown in Fig. 3. To start the fault tree derivation, it is at first recognized that the top event corresponds to the node $F=10.0$. Backtracking through the model (Fig.2), one finds that the inputs H and VX are associated through the transfer box A to the output F. In the column corresponding to F in Table 2, only one entry element is found that has the equal to 10.0. The H-column and VX-column entries in the same row of this F-column entry, $H=5$ and $VX=1.0$, cause $F=10.0$. This information is represented in the fault tree using an AND gate with events $H=5$ and $VX=1.0$ as its inputs. It should also be noted that $H=5$ is not developed any further, since it is a basic event and hence is represented by a circle. The $VX=1.0$ is developed further, since it is the output of transfer box C.

72

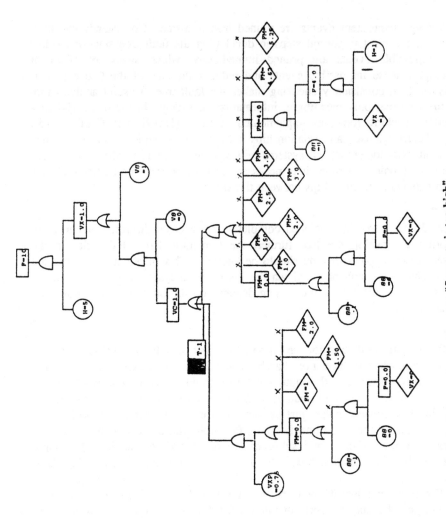

Fig.3. Timed fault tree for system state "flow rate is too high"

Looking at the fault tree, it can be seen that all events in the fault tree have been developed until they lead to either a basic event or an output from a transition box at the current time. Here $VC=1.0$ is the only event that is an output of a transition box (DCMB) and, hence there is no stochastic indetermination as to which transition may have occurred last. A flag, which is shown as a rectangle half filled with black, has been placed below the event $VC=1.0$ to indicate that time transition DCMB has occurred and that, below that tree branch, the system is in the preceding time frame. The notation "$T-1$" in the flag in Fig.3 indicates that the system is in the cycle-before-last in the backtracking process. Now $VC=1.0$ can be developed further and the inputs FM and VXP that cause $VC=1.0$ are found from Table 2 and are ordered in the fault tree using AND and OR gates. These procedures are continued until the timed fault tree shown in Fig. 3, is fully developed.

Note that two consistency checking rules have been applied to the fault tree in our example in order to prune branches that were not consistent with already existing events in the fault tree. These are:

1. The height of water in the tank cannot be greater than in the previous cycle, since there is no source for replenishing the water draining out of the tank. For example, in the fault tree shown in Fig.3, $H=5$ and $VX=1.0$ cause the top event $F=10$. Developing $VX=1.0$ through $FM=4.0$, we obtain $H=1$ in the previous cycle. Since $H=5$ has occurred in present cycle, then $H=1$ is not a possible event in the previous cycle. Therefore, the branch $H=1$ should be pruned from the tree. Since events above $H=1$ in the fault tree (e.g., $F=4.0$ and $FM=4.0$) can occur only if $H=1$ occurs (because they are connected to $H=1$ through an AND gate), they are inconsistent and, hence, are also pruned from the tree.

2. The value of VXP (valve position in the previous cycle) obtained while back-tracking in the present cycle should be the same as the value of VX (valve position at the present time) obtained while backtracking in the previous cycle. For example, in the fault tree shown in Fig.3, when developing event $VC=1.0$, we obtain $VXP=0.75$ and $FM=0.0$. This means that the valve position in the previous cycle must be 0.75. But, while developing $FM=0.0$ further, we obtain the event $VX=0$ instead. Hence, $VX=0$ is not possible event and is pruned from the tree.

In the fault tree shown in Fig.3, all the branches marked by "×" were found to be inconsistent and were pruned from the fault tree. The analysis of the final fault tree is discussed in Section 4.

4 Analysis of Timed Fault-Trees

A timed fault tree, which is a logical combination of snapshots of system evolution, provides valuable documented information about the nature and behavior of the system. The most commonly derived results from fault trees are the minimal cut-sets for the top event. Each minimal cut-set is a complete set of the minimum necessary and sufficient combinations of hardware and software state(s) that can potentially cause the top event. Traditional fault-tree analysis with binary state components uses boolean algebra to reduce the fault tree in order to obtain minimal cut-sets. An upper limit on the order of the cut-set (i.e., number of simultaneous basic events) is usually set so that only the most important minimal cut-sets are determined. But, in order to derive the minimal cut-sets for DFM timed fault trees, special methods are required because DFM uses a multi-state (i.e. more than two states) representation of variables and this translates into multi-valued, rather than binary, fault trees. One such method is outlined in [28]. This method is useful to derive minimal cut-sets from a fault tree developed for a system which can be described by multi-state variables.

Two minimal cut-sets for the fault tree shown in Fig. 3 are:

$$(H = 5.0)*(VS = 1) +$$

$$(H = 5.0)*(VS = 0)*(VXP = 0.75 + VXP = 1.0)*(SS = -1).$$

For example, the first minimal cut-set for the top event "flow-rate is too high" is: the valve is fully open, the height of the water column is 5.0 m, and that the time taken to reach the top event from the initial state of the system is one cycle for the sequence of events represented by this cut-set. The second minimal cut-set shows that the time taken to reach the top event from the initial state of the system is two cycles, with $SS=-1$ occurring in the last-but-one cycle and the remaining events in the cut-set occurring in the last cycle. The flag that is placed in the fault tree to show DCMB has occurred (Fig. 3) indicates that all events occurring below that branch and above the flag corresponding to the previous time transition, happened in the last-but-one cycle. In fact, these flags are the mechanism by which software and hardware nodes are time-stamped in the fault tree.

5 Conclusions

The dynamic flowgraph methodology (DFM) provides a relatively simple and effective method to represent and analyze an embedded system. The model includes important time elements and causal relationships that describe the system. It is used to build fault trees for identified system states of interest by using multi-state variables. A notable feature that distinguishes DFM from other methods is that separate model constructions are not necessary for every system state of interest. The "timed" fault trees, derived from the model, give information on the time sequence of software and hardware states that can cause the system to be in a certain state. The systematic backtracking through the model to build "timed" fault trees lends itself readily to automated construction and could be a very useful design tool. It should be noted that, in the construction of timed fault trees, since one needs only the reachability graph of transition boxes, a state transition network can be used instead of a Petri network. This could possibly save some significant amount of time and effort in analyzing systems of a certain complexity.

The information about faulted states of hardware and software that cause a system to be in a certain state can be used advantageously during embedded system design and review to make the system more dependable (by eliminating uncovered design errors or unanticipated software--hardware interactions), or to develop more directed system-specific testing strategies. Testing usually uncovers almost all failures caused by the state of a single node (single point failures), but it has been less successful in uncovering those caused by coupled states. The minimal cut-sets provide information on the coupled states that can cause the undesirable top event, and this, in conjunction with testing, can be used to verify that these system failure modes are not freely active in the system. If A and B can cause the top event and A is an external event whose occurrence cannot be controlled, then one can test the system by allowing A to occur (i.e. making the software "believe" that it has) and test whether B can be induced to occur by the software itself. We call this sequence "dormant logic and trigger" events, where occurrence of B is dormant in the logic of the system and the top event is triggered by occurrence of external event A. To improve the dependability of the system, safeguards have to be built into it to prevent the simultaneous occurrence of these event combinations.

The timed fault trees act as a testing ground to check whether the embedded system, which comprises the various components of hardware and software, performs its function properly and, hence, complements integration testing. These "timed" fault trees provide one single structure for identifying failures in the software, hardware, and their interfaces. DFM analyzes the system as a whole and, thereby, provides an effective framework to test its dependability features.

Acknowledgements

This paper was supported by a grant from NASA Goddard Space Flight Center and by a grant from the UCLA Flight Systems Research Laboratory sponsored by NASA Ames Research Center.

References

1. Laprie, J. C.: Dependability: a unifying concept for reliable computing and fault-tolerance, Resilient Computing Systems. London, Collins, 1-28 (1989)

2. Ramamoorthy, C. V. et al: Application of methodology for the development and validation of reliable process control software. IEEE Trans. Software Eng. 7, 537-555 (1981)

3. Leveson, N. G., Stolzy, J. L.: Safety analysis using Petri nets. IEEE Trans. Software Engineering, 13, 386-397 (1985)

4. Leveson, N. G.: Software safety in computer controlled systems," IEEE Computer, 17, 48-55 (1984)

5. Leveson, N. G.: Software safety: why, what and how. ACM Computing Surveys, 18, 125-163 (1986)

6. McDermid, J. A.: Issues in developing software for safety critical systems. Reliab. Engng. & System Safety, 32, 1-24 (1991)

7. Parnas, D. L., Asmis, G. J. K., Madey, J.: Assessment of safety-critical software in nuclear power plants. Nuclear Safety, 32, 189-198 (1991)

8. Harvey, P. R.: Fault-tree analysis of software. Master's thesis, Univ. California, Irvine (1982)

9. Leveson, N. G., Harvey, P. R.: Analyzing software safety. IEEE Trans. Software Engineering, SE-9, 569-579 (1983).

10. DeMillo, McCracken, et al, "Software Testing and Evaluation", Benjamin/Cummings Publishing Company Inc., Menlo Park, CA,1987.

11. Peterson, J. L.: Petri net theory and the modeling of systems. Prentice-Hall, Inc., Englewood Cliffs, NJ (1983)

12. Jones, C. B.: Software development: A rigorous approach," Prentice-Hall Inc., Englewood Cliffs, NJ (1986)

13. Mills, H. D.,Dyer, M., Linger, R.: Clean room software engineering. IEEE Software, 4, 19-25 (1987)

14. Lapp, S. A., Powers, G. J.: Computer-aided synthesis of fault trees. IEEE Trans. Reliability, 26, 2-13 (1977)

15. Salem, S. L., Apostolakis, G. E.: A new methodology for the computer-aided construction of fault trees. Annals of Nuclear Energy, 4, 417-433 (1977)

16. Salem, S. L., Wu, J. S., Apostolakis, G. E.: Decision table development and application to the construction of fault trees. Nucl. Technol., 42, 51-64, 1979.

17. Han, S. H., Kim, T. W., Choi, Y.: Development of a computer code AFTC for fault tree construction using decision tables and supercomponent concept. Reliab. Engng. & System Safety, 25, 15-31 (1989)

18. Guarro, S. B., Okrent, D.: The Logic Flowgraph: a new approach to process failure modeling and diagnosis for disturbance analysis application. Nucl. Technol., 67 (1984)

19. Guarro, S. B.: PROLGRAF-B: a knowledge-based system for the automated construction of nuclear plant diagnostic models. In technical progress report for period Sep. 1987-March 1988 (by D. Okrent and G. Apostolakis) for DOE award no. DE-FGO3-UCLA, March 1988.

20 Guarro, S. B.: Diagnostic models for engineering process management: A critical review of objectives, constraints and applicable tools. Reliab. Engng. & System Safety, 30, (1990)

21. Muthukumar, C. T., Guarro, S. B., Apostolakis, G. E.: Logic Flowgraph Methodology: A tool for modeling embedded systems. IEEE/AIAA 10th Digital Avionics Systems Conference, Los Angeles, CA, Oct. 14-17. Proceedings, 103-107 (1991)

22. Amendola, A.: Accident sequence dynamic simulation versus event trees. Reliab. Engng. & System Safety, 22, 3-25 (1988).

23. Nivoliantou, Z., Amendola, A., Reina, G.: Reliability analysis of chemical processes by the DYLAM approach. Reliability Engineering, 14, 163-182 (1986)

24. Cacciabue, P. C., Amendola, A.: Dynamic Logical Analytical Methodology versus fault trees: The case study of the auxiliary feedwater system of a nuclear power plant. Nucl. Technol., 74 (1986).

25. Taylor, J. R.: An algorithm for fault-tree construction, IEEE Trans. Reliability, 31, 137-146 (1982)

26. Kelly, B. E., Lees, F. P.: The propagation of faults in process plants. Reliability Engineering, 16, 1-35 (1986)

27. Coolahan, J., E. Jr., Roussopoulos, N.: Timing requirements for time-driven systems using augmented Petri nets. IEEE Trans. Software Engineering, SE-9, 603-616 (1983).

28. Caldarola, L.: Fault tree analysis with multistate components, Synthesis and Analysis Methods for Safety and Reliability Studies. Apostolakis, G. E., Garribba S., Volta, G. (Eds), 199-248, New York: Plenum Press (1980).

Part 2

Dynamic Approaches - Applications

Dynamic Programming and Optimization

Dynamic Approaches - Applications: An Overview

Tunc Aldemir

Nuclear Engineering Program, Department of Mechanical Engineering,
The Ohio State University, Columbus, OH 43210, USA

Abstract. Some applications of dynamic methodologies for process system reliability and safety analyses are described and the advantages and limitations of these methodologies are discussed.

Keywords. Dynamic methodologies, Markov models, dynamic event tree approach, direct system simulation

1 Introduction

A survey of the literature [1,2] shows that dynamic methodologies for process system reliability and safety analyses can be grouped under three categories:

- State transition or Markov models
- Dynamic event tree approach
- Direct system simulation or analog Monte Carlo

This paper gives some examples of the applications of dynamic methodologies, as well as their advantages and limitations as perceived by the author.

2 Markov Models

Markov models represent the time evolution of the system as possible transitions between system states. The transition probabilities should not depend on the history of system operation. The model yields the probability of finding the system at a given state and at a given time. Markov models are most commonly used in reliability and safety analyses when the system states are discrete and the transitions between the system states are statistically dependent. Two examples of such applications are given by Walker [3] and Dhillon [4] where the system states can be described purely in terms of the hardware and operator states. Devooght and Smidts [5] show that, using a Chapman-Kolmogorov interpretation of system dynamics, Markov modeling techniques can be extended to systems whose state-space includes both process variables with continuous states and components with discrete states. A variation of this technique that uses discrete process variable states (i.e. magnitude intervals) has been independently developed and applied to a simple tank system [6] and also to determine the reliability of the feed-bleed cooling of a boiling water reactor core during 60 minutes following a small-break which incapacitates the reactor core isolation cooling system [7].

Markov models can handle:

- time-dependent system parameters,

- operator behavior dependent on the current hardware and process variable states or changes in the process variable states, and,

- statistical dependencies arising from history independent interactions of hardware-hardware, hardware-operator, hardware-process variable, control law-process variable changes.

Modeling the statistical dependencies between the control laws and process variable changes may require the discrete process variable representation scheme. This representation scheme is also useful for utilization of failure-per-demand data as reported, however, suffers worse than some other approaches regarding limitations on system size due to computational limitations. The other difficulty with Markov models is that operator and/or hardware behavior dependent on system history cannot be represented.

3 Dynamic Event Tree Approach

The dynamic event tree approach tracks possible branchings in system evolution at specified time intervals following an initiating event through simulations. The simulations stop when a specified number of time intervals or a Top Event is reached. The sum of scenario (event sequence) probabilities leading to a Top Event gives the probability that this Top Event will occur during the specified time interval. DYLAM is the first computational tool that has been developed using this approach [8]. Izquierdo et al [9] show how DYLAM can be coupled to a generic simulation tool for nuclear power plant analysis. Two variations of DYLAM are DETAM [10] and ADMIRA [11]. DETAM allows a more general treatment of stochastic variability in operator behavior and ADMIRA allows on-line modification of plant status. An application of DETAM to a pressurized water reactor steam generator tube rupture is given in [10].

The techniques using dynamic event trees have almost all the capabilities of the techniques using Markov models. One computational advantage of these techniques is that, if the future system behavior does not depend on the history of system operation, they require very little computer storage compared to Markov models. However, such applications require the development of very fast simulators and analyses have to be repeated every time component failure data are changed. Also, for the representation of history dependent operator or component behavior, all generated event sequences need to be saved. In these situations, the dynamic event tree approach also becomes very storage-limited.

4 Direct System Simulation

In contrast to the dynamic event tree approach, direct system simulation techniques do not track all possible pre-specified scenarios but rather use the Monte Carlo approach of building a pattern of system responses to an initiating event and determining the probability of specified outcomes by sampling from the infinite number of possible branchings. Using this technique on the tank system of [6], Marseguerra [12] has shown that when the reliability characteristics of the plant strongly depend on the process, the results obtained from direct system simulation and the classical event-tree/fault-tree approach are quite different. Another example to direct system simulation is the failure analysis of the decay heat removal system of a liquid metal fast breeder reactor by Nakada and his co-workers [13]. For this system, the failure rates are time and event sequence dependent. Direct system simulation has been also applied to obtain estimates of the probability of human error [14].

Almost all types of hardware-operator-process variable interactions of interest and time/process variable/event sequence dependent failure rates can be modeled using direct simulation techniques. They have much lower computer storage requirements compared to Markov models if the definition of system states is based on component states and discrete process variable states. Also, they may require less computer time compared to the dynamic event tree approaches with proper variance reduction techniques. However, variance reduction techniques are usually problem specific and complicate the model. Also, Monte Carlo techniques may not sample from rare event sequences and subsequently miss their consequences. Another problem is that the results give the integrated effect of the event sequences simulated so it is usually very difficult to obtain information about the individual event sequences leading to a given Top Event. Finally, the analysis has to be repeated if the failure data are changed so they may not be suitable for sensitivity analyses.

5 Conclusion

Applications to date show that dynamic methodologies are usually needed whenever there is complex hardware-operator-process variable interaction in time. It is hard to identify one dynamic methodology as clearly superior to the rest. Each methodology has advantages and disadvantages depending on the system under consideration and the objectives of the analysis. However, they all suffer from computational limitations on system size and are usually used in conjunction with some system size reduction technique such as definition of macrocomponents through failure-modes-and-effects analysis or state aggregation [15].

References

1. Siu, N. O.: Risk assessment for dynamic systems: an overview. Accepted for publication in Reliab. Engng. & System Safety.

2. Siu. N. O.: Dynamic approaches - issues and methods: an overview. These proceedings.

3. Walker, B. K.: Evaluating performance and reliability of automatically reconfigurable systems using Markov modeling techniques. These proceedings.

4. Dhillon, B. S.: Reliability analysis under fluctuating environment using Markov method. These proceedings

5. Devooght, J., Smidts, C.: Probabilistic dynamics: the mathematical and computing problems ahead. These proceedings

6. Aldemir, T: Computer-assisted Markov failure modeling of process control systems. IEEE Trans. Reliability, R-36, 133-144 (1987)

7. Hassan, H., Aldemir, T.: A data base oriented dynamic methodology for the failure analysis of closed loop control systems in process plants. Reliab. Engng. & System Safety, 27, 275-322 (1990)

8. Cacciabue, P. C., Carpignano, A., Cojazzi, G.: Dynamic reliability analysis by DYLAM methodology: the case of CVCS of a PWR. Probabilistic Safety Assessment and Management, G. Apostolakis (Ed.), 539-544, New York: Elsevier (1991)

9. Izquierdo, J. M., Hortal, J., Sanchez, M., Melendez, E.: Automatic generation of event trees: A tool for safety assessment. These proceedings

10. Siu, N., Acosta, C.: Dynamic event tree analysis-An application to STGR. Probabilistic Safety Assessment and Management, G. Apostolakis (Ed.), 413-418, New York: Elsevier (1991)

11. Senni, S., Semenza, M. G., Galvani, R.: A.D.M.I.R.A. - An analytical dynamic methodology for integrated risk assessment. Probabilistic Safety Assessment and Management, Apostolakis, G. (Ed.), 407-412, New York: Elsevier (1991)

12. Marseguerra, M., Zio, E.: Approaching dynamic reliability by Monte Carlo simulation. These proceedings

13. Nakada, K., Miyagi, K., Handa, N., Hattori S.: A method of state transition analysis under system interactions: An analysis of a shutdown heat removal system. Nucl. Technol., 82, 132-146 (1989)

14. Vestrucci, P., Santrucci, R., Calderan, R.: Monte-Carlo simulation of crew responses to accident sequences. Reliab. Engng & System Safety, 31, 129-144 (1991)

15. Papazoglou, I. A., Gyftopulos, E. P.: Markov processes for reliability analysis of large systems. IEEE Trans. Reliability, R-26, 232-237 (1977).

Probabilistic Dynamics : The Mathematical and Computing Problems Ahead

Jacques Devooght[1] and Carol Smidts[1,2]

[1] Université Libre de Bruxelles, Brussels, Belgium.
[2] On leave at JRC/Ispra (ISEI), Italy.

Abstract. The methodology of probabilistic dynamics viewed as a continuous event tree theory is reviewed and other existing methods are shown to be particular cases corresponding to definite assumptions. Prospects for improvement of related numerical algorithms are examined.

Keywords. Probabilistic dynamics, continuous event tree, markovian reliability.

1 Probabilistic Dynamics

"Probabilistic dynamics" is "system dynamics" supplemented by the fact that deterministic trajectories in phase space switch at random times to other trajectories, either because of stochastic changes in the structure of the system (for instance change of state of components) by failure, or built in control, or still because of human action. In order to include the operator's behaviour we must extend the meaning of the "system" which involves coupling between the physical system and the operator [1,2].

Variables describing the physical system appear in a vector $\bar{x}$, with components like average temperature, average power, reservoir level, etc. Phase space R^N with $\bar{x} = (x_1...x_N)$ may be in practice, for realistic systems, of dimensions large enough to preclude conventional means of solution. The structure of the system may be characterized by the state of the components and the usual setting is a markovian model with m^n states where n is the number of components, each being in m possible states. The operator "state" may be characterized in a similar way, although the way to quantify its "state of mind", stress level, etc., is a delicate and barely explored subject. If we assume a total of M states for the operator we have a total of Mm^n discrete spaces and the space into which we must work is Mm^n replicas of R^N.

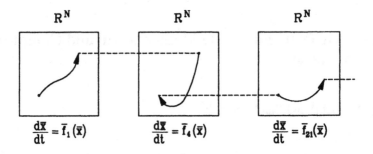

$$\frac{d\bar{x}}{dt} = \bar{f}_1(\bar{x}) \qquad\qquad \frac{d\bar{x}}{dt} = \bar{f}_4(\bar{x}) \qquad\qquad \frac{d\bar{x}}{dt} = \bar{f}_{21}(\bar{x})$$

For instance in the figure above we have a trajectory in the initial discrete state 1 which jumps to another trajectory in the 4th state, which jumps again to a trajectory in the 21st state, etc.

The deterministic dynamics is characterized by the set of differential equations

$$\frac{d\bar{x}}{dt} = \bar{f}_i(\bar{x}) \tag{1}$$

where index i labels the discrete component or operator states. Hopefully not all $\bar{f}_i$ are different.

The history of a transient in a system like a nuclear reactor-operator system is a succession of states $(\bar{x}_1,i_1,t_1), (\bar{x}_2,i_2,t_2)...(\bar{x}_k,i_k,t_k)$ where a transition at time t_k is in state $\bar{x}_k(t_k)$ and has just entered discrete state i_k. Our objective is to compute the probability density $\pi(\bar{x},i,t|\bar{x}_o,k,t_o)$ where $(\bar{x}_o,k,t_o)$ is the initial state. The mathematical theory needed is well known at least for markovian systems. Obviously the deterministic trajectory is markovian since $\bar{x}(t+\Delta t)$ is determined completely and uniquely by $\bar{x}(t)$. The markovian assumption is good enough for components where even aging effects can be taken into account either by explicit time dependence or by additional fictitious states. The markovian assumption for operators is open to question but an extension to semi-markovian models does not complicate the actual numerical calculation while it allows time distributions which are general and not only exponential. We cannot show here for lack of space how to deal with semi-markovian modelling [37,38].

It can be shown [1] that $\pi(\bar{x},i,t|\bar{x}_o,k,t_o)$ obeys a differential Chapman-Kolmogorov [3] equation, in the case of a markovian assumption :

$$\frac{\partial}{\partial t}\pi(\bar{x},i,t|\bar{x}_o,k,t_o) + div_x[\bar{f}_i(\bar{x})\pi(\bar{x},i,t|\bar{x}_o,k,t_o)]$$

$$+ \lambda_i(\bar{x})\pi(\bar{x},i,t|\bar{x}_o,k,t_o) - \sum_{j\neq i} p(j\rightarrow i|\bar{x})\pi(\bar{x},j,t|\bar{x}_o,k,t_o) = 0 \tag{2}$$

We shall usually write $\pi(\bar{x},i,t)$ for short.

The transition rate $p(j\rightarrow i|\bar{x})$ may be dependent upon the current state vector. This dependence is essential because it embodies not only the fact that the failure rate may be, for instance, temperature dependent, but also that valves open or close if pressure crosses a threshold level, or operators act on the cue of a signal, etc. [4,5].

We have finally

$$\lambda_j(\bar{x}) = \sum_{i \neq j} p(j \rightarrow i \,|\, \bar{x}) \tag{3}$$

Equation (2) yields exponential holding time distributions in each state and is completely defined by $\bar{f}_i(\bar{x})$ and $p(j \rightarrow i \,|\, \bar{x})$. We shall need the solution of eq. (1) with

$$\bar{x}(t) = \bar{g}_i(t, \bar{x}_o) \ , \quad \text{with} \quad \bar{x}(o) = \bar{g}_i(o, \bar{x}_o) = \bar{x}_o$$

Transforming eq. (2) into an integral form [1], we obtain :

$$\lambda_i(\bar{x})\pi(\bar{x}, i, t) = \lambda_i(\bar{x}) \int \pi(\bar{u}, i, t_o) \delta(\bar{x} - \bar{g}_i(t, \bar{u}))(1 - F_i(t, \bar{u})) d\bar{u}$$

$$+ \sum_{j \neq i} \int_{t_o}^{t} \int \lambda_j(\bar{u})\pi(\bar{u}, j, t - \tau) \frac{p(j \rightarrow i \,|\, \bar{u})}{\lambda_j(\bar{u})} \delta(\bar{x} - \bar{g}_i(\tau, \bar{u})) dF_i(\tau, \bar{u}) d\bar{u} \tag{4}$$

where

$$F_i(t, \bar{u}) = 1 - e^{-\int_{t_o}^{t} \lambda_i[\bar{g}_i(s, \bar{u})] ds} \tag{5}$$

is the probability that the system leaves state i before t, during its deterministic drift. The rate of transition $\lambda_i(\bar{x})\pi(\bar{x}, i, t)$ out of state i is the sum of two terms : the first corresponds to the case where the system has never left the initial state i, the second to transitions from $(\bar{u}, j)$ occurring at $t - \tau$, leading to i with a conditional probability $\dfrac{p(j \rightarrow i \,|\, \bar{u})}{\lambda_j(\bar{u})}$ followed by a deterministic drift along $\bar{g}_i(\tau, \bar{u})$ for a duration τ and finally having a transition out of i at time t. The appearance of the delta function $\delta(\bar{x} - \bar{g}_i(t, \bar{u}))$ is natural since in state i, the evolution is deterministic.

We can now release our markovian assumption eq. (5) and allow for any CdF of the transition time, giving an extended meaning to eq. (4). Each discrete state is assumed however to be a renewal state [6]: the time distribution is non exponential (semi-) although the probability $p(j \rightarrow i \,|\, \bar{x})$ depends only on (i,j) (-markovian). Equation (4) describes a transport process in phase space where a complete correspondence with the Boltzmann equation for neutron transport can be established with however an important difference : trajectories are curves $\bar{x} = \bar{g}_i(t, \bar{u})$ in N dimensional space and not straight lines in 3 dimensional space! The state index i plays the same role as angle or group index.

We show now the relevance of eqs. (1)-(5) to safety problems, or in general to dynamic reliability problems as summarized in [7]. A safety incident or accident

88

is characterized by the fact that process variables cross a threshold. The set of all thresholds defines a safety boundary which is usually an hypercube, or a simplex if some trade off is allowed between variables. The classical method of defining reliability by means of cut sets is not necessarily relevant here because these cutsets are usually defined after consideration of the transients they induce. Safety belongs therefore to the category of exit problems studied in other fields of physics and we want to know the probability that a transient will exit, given its initial condition, or top event which defines the accident. We may partition the safety boundary - defined by the safety analyst in function of its objectives - into $\Gamma = \Gamma_+(i) \cup \Gamma_-(i)$ for each i, with the outgoing boundary $\Gamma_+(i)$ defined by $\overline{n}\overline{f}_i(\overline{x}_\Gamma) > 0$ where $\overline{n}$ is the outgoing normal at point $\overline{x}_\Gamma \in \Gamma(i)$. If we impose a boundary condition like $\pi(\overline{x}_\Gamma, i, t | \overline{x}_o, k, t_o) = 0$ for $\overline{x}_\Gamma \in \Gamma_-(i)$, we assume that all outgoing trajectories are "absorbed" which means that in fact we study the distribution of the first exit time [8,9].

Among byproducts of eq. (2) we have for instance the system of partial differential equations for the mean escape time at any point Γ if we start in state $(\overline{x},i)$. Function $MTTF(\overline{x},i)$ generalizes the usual mean time to failure used in classical markovian reliability :

$$MTTF(\overline{x},i) = \frac{1}{\lambda_i(x)} + \sum_{k \neq i} \frac{p(i \to k | \overline{x})}{\lambda_i(x)} MTTF(\overline{x},k) + \frac{\overline{f}_i(\overline{x})\overline{\nabla}MTTF(\overline{x},i)}{\lambda_i(x)} \qquad (6)$$

We recognize the standard form with the exception of the last term which expresses the fact that while system is in state i, the system drifts to an neighboring point in phase space where $MTTF(\overline{x},i)$ is different. If $\overline{f}_i = 0$ we recover the usual result [6].

2 Analysis of Other Methods Derived From the General Case

Before examining methods of solution of eqs. (1) and (2), we must first show how various other methods developed by other authors in these last years can be obtained as special cases. We do it only on general terms, to the best of our knowledge and we refer the reader to the literature and other papers in these proceedings for full details. Our analysis, with our notation, deals only with the mathematical aspects.

2.1 The Pure Markovian Approach

If we integrate eq. (2) over all $\overline{x}$ space, we obtain an equation of the type :

$$\frac{d\pi(i,t)}{dt} = -<\lambda_i>\pi(i,t) + \sum_{j\neq i} <p(j\rightarrow i)>\pi(j,t) \qquad (7)$$

with $\pi(i,t) = \int \pi(\bar{u},i,t)\,d\bar{u}$.

The divergence term disappears because $\lim\limits_{|\bar{x}|\rightarrow\infty} \pi(\bar{x},i,t)=0$.

We recover the standard markovian model used in reliability theory [11,12] if $p(j\rightarrow i|\bar{u})=p(j\rightarrow i)$, because then we do not need the full $\pi(\bar{u},j,t)$ distribution.

2.2 State Variable Discretization

To reduce dimensionality we may have to discretize $\bar{x}$ or t variables in a sense explained below, or aggregate states in groups I with $i \in I$.

Let us first divide R^N into disjoint cells $D_k: R^N = \underset{k}{U} D_k$.

If $\pi_k(i,t) = \int_{D_k} \pi(\bar{x},i,t)d\bar{x}$, and $H_k(\bar{x})$ the characteristic function of cell k defined by

$$H_k(\bar{x})=1 \text{ if } \bar{x} \in D_k$$
$$H_k(\bar{x})=0 \text{ if } \bar{x} \notin D_k .$$

Then integration of eq. (4) gives, after division by $\lambda_i(\bar{x})$, and in the full markovian form

$$\pi_k(i,t) = \int \pi(\bar{u},i,t_o) H_k[\bar{g}_i(t,\bar{u})] e^{-\int_{t_o}^t \lambda_i[\bar{g}_i(s,\bar{u})]ds} d\bar{u}$$

$$+ \sum_{j\neq i} \int d\bar{u} \int_{t_o}^t d\tau \, \pi(\bar{u},j,t-\tau) H_k[\bar{g}_i(\tau,\bar{u})] p(j\rightarrow i|\bar{u}) e^{-\int_{t_o}^t \lambda_i[\bar{g}_i(s,\bar{u})]ds} \qquad (8)$$

So far no approximation is made. If cells are chosen such that discontinuities of $p(j\rightarrow i|\bar{u})$ appear only on the cell's boundaries, $p(j\rightarrow i|\bar{u})$ will be fairly regular and can be approximated for instance by $p(j\rightarrow i|\bar{u}_\ell)$ where $\bar{u}_\ell$ is the "center" of cell D_ℓ.

However $H_k[\bar{g}_i(t,u)]$ may vary more rapidly over the domain : let us replace H_k by its interpolation $\hat{H}_k$, constant over each cell of volume V_ℓ.

$$\hat{H}_k[\bar{g}_i(t,\bar{u})] = \sum_\ell H_\ell(\bar{u}) \int_{D_\ell} H_k[\bar{g}_i(t,\bar{w})] \frac{d\bar{w}}{V_\ell} \equiv \sum_\ell H_\ell(\bar{u}) H_{k\ell}(i,t) \qquad (9)$$

If we choose a small increment $t=\Delta t$ developing $\pi_k(i,\Delta t)$ to first order and defining a transition probability from cell ℓ to cell k

$$p(\ell \rightarrow kli) \;=\; \lim_{\Delta t \rightarrow o} \frac{H_{k\ell}(i,\Delta t) - \delta_{k\ell}}{\Delta t} \tag{10}$$

we obtain essentially the result of Aldemir et al. [13-17]. We remark that we have a markovian problem with double index (i,k) states and it can be shown easily that we obtain the markovian evolution equation

$$\frac{d\pi_k(i,t)}{dt} \;=\; -[\lambda_i(\bar{u}_k) + \sum_{\ell \neq k} p(k \rightarrow \ell li)]\,\pi_k(i,t)$$

$$+ \sum_{j \neq i} \pi_k(j,t)\,p(j \rightarrow i|\bar{u}_k) + \sum_{\ell \neq k} \pi_\ell(i,t)\,p(\ell \rightarrow kli) \tag{11}$$

Movement in discretized R^N space is pseudo probabilistic because trajectories issued from neighboring points of the same cell may end, at Δt, in different cells [18]. We have in fact a self transition $p(k \rightarrow kli)$ to the same cell and $H_{k\ell}(i,t)$ has the following meaning : if starting points of trajectories in state i are uniformly distributed over cell ℓ, the proportion of trajectories arriving in cell k at time t is $H_{k\ell}(i,t)$.

2.3 Time Discretization

We have an alternate discretization, of the t variable, which leads to event simulation methods like DYLAM [19-22] or DETAM [7,23].

Let us imagine that instead of having a CdF $F=1-e^{-\lambda t}$, we have an approximate $F^a(t) = \sum_n q_n H(t-t_n)$, with $\sum_n q_n = 1$. The staircase function $F^a(t)$ must approximate $F(t)$ by an appropriate choice of q_n, the probability that a transition will occur at the predetermined time t_n.

Let $\pi^{(n)}(\bar{x},i,t)$ be the distribution after n transitions :

$$\pi(\bar{x},i,t) \;=\; \sum_{n=o}^{\infty} \pi^{(n)}(\bar{x},i,t) \quad . \tag{12}$$

If we substitute

$$F_i^a(t,\bar{u}) = \sum_n q_n^i H(t-t_n^i(\bar{u})) \tag{13}$$

for $F_i(t,\bar{u})$ and take, for the sake of illustration $\pi(\bar{u},i,t_o) = \delta_1^i \delta(\bar{u}-\bar{x}_o)$

$$\pi^{(o)}(\bar{x},i,t) = \delta_1^i \,\delta(\bar{x}-\bar{g}_1(t,\bar{x}_o))(1-F_1(t,\bar{x}_o)) \tag{14}$$

The staircase behaviour of $F^a(t)$ means that transitions occur instantly at a given time. We must reformulate in terms of rate of transitions <u>out</u> of state i : $R(\bar{x},i,t) \equiv \lambda_i(\bar{x})\,\pi(\bar{x},i,t)$ which is

$$R(\bar{x},i,t) = R_o(\bar{x},i,t) + \sum_{j\neq i} \int\int_{t_o}^t R(\bar{u},j,t-\tau) \frac{p(j\to i|\bar{u})}{\lambda_j(u)} \delta(\bar{x}-\bar{g}_i(\tau,\bar{u})) dF_i(\tau,\bar{u}) d\bar{u} \quad (15)$$

where $R_o(\bar{x},i,t)$ is the rate of leaving the initial state.

Let $R^{(n)}(\bar{x},i,t)$ be the rate of transitions if n previous transitions have already occurred. Then :

$$R^{(n)}(\bar{x},i,t) = \delta_o^n R_o(\bar{x},i,t) + \sum_{j\neq i} \int d\bar{u} \int_{t_o}^t R^{(n-1)}(\bar{u},j,t-\tau) \frac{p(j\to i|\bar{u})}{\lambda_j(u)} \delta(\bar{x}-\bar{g}_i(\tau,\bar{u})) dF_i(\tau,\bar{u})$$

$$(16)$$

with

$$R_o(\bar{x},i,t) = -\frac{d\pi^{(o)}}{dt}(\bar{x},i,t) = \delta_1^i \sum_n \delta(\bar{x}-\bar{g}_i(t,\bar{x}_o)) q_n^1 \delta(t-t_n^1) \quad (17)$$

and

$$R^{(1)}(\bar{x},i,t) = \sum_{n,m} q_n^1 \delta(\bar{x}-\bar{g}_i(t_m^i,\bar{g}_1(t_n^1,\bar{x}_o))) q_m^i \delta(t-t_n^1-t_m^i) \hat{p}(1\to i|\bar{g}_1(t_n^1,\bar{x}_o)), \quad (18)$$

the rate λ_i not appearing explicitly since we need only the conditional probability
$$\hat{p}(j\to i|\bar{u}) = p(j\to i|\bar{u})/\lambda_j(\bar{u}) \quad .$$

We observe that the second transitions are fired at times $t_n^1 + t_m^i$ and at this time the representative point of the system is at

$$\bar{x} = \bar{g}_i(t_m^i,\bar{g}_1(t_n^1,\bar{x}_o))$$

which is indeed the position if we go from $\bar{x}_o$ to $\bar{g}_1(t_n^1,\bar{x}_o)$ at time t_n^1, and then go from this point to reach t_m^i where a second transition occurs. The probability is the product of the probability that the first transition occurs at t_n^1 times the conditional probability that if the first state is 1, the second is i, times the probability that the second transition occurs at $t_n^1 + t_m^i$.

The complexity of the event tree is already well apparent, particularly if the tree is asynchronous, i.e. the t_n^i are not multiples of a common Δt for all i. What error is produced by the fact that we use an approximate F_i^a instead of the exact one? It can be shown that the $\| \ \|_1$ norm of the error is bounded by

$$\|\pi(\bar{x},i,t) - \pi^a(\bar{x},i,t)\|_1 \equiv \sum_i \int |\pi(\bar{x},i,t) - \pi^a(\bar{x},i,t)| d\bar{x} \leq |\Delta F_1(t)| + \underset{j\neq 1}{Max} |\Delta F_j(\xi)|$$

with $\qquad\qquad\qquad 0 \leq \xi \leq t \qquad\qquad\qquad\qquad\qquad (19)$

where

$$\Delta F_i(t) = \underset{\bar{u}}{Max} |F_i(t,\bar{u}) - F_i^a(t,\bar{u})| \quad . \tag{20}$$

Obviously $\|\pi-\pi^a\|_1 \to 0$ if $|\Delta F_i(t)| \to 0$, which means that the event tree should of course be endlessly refined with more and more transitions. To attempt an estimate of the error we must compare with an equivalent Monte-Carlo treatment. One way to test if trajectories obtained by $F^a(t)$ are a statistical sample of $F(t)$ is to use the Kolmogorov-Smirnov test [35] : if $\underset{t}{Max}|F^a(t)-F(t)|<D_n^\gamma$ there is no statistically significant difference between the distributions at the γ level for a sample of size n. For $\gamma = 0.05$, and n > 35, we may write $D_n^\gamma \sim 1.36/\sqrt{n}$.

If we want to approximate $F(t) = 1 - e^{-\lambda t}$ by $F^a(t) = \dfrac{1}{n}\sum_{k=1}^{n} H(t-t_k)$ the criterion is certainly respected if the maximum differences are all equal, i.e.

$$e^{-\lambda t_k} + e^{-\lambda t_{k+1}} = \frac{2n-2k+1}{2n} \tag{21}$$

with k=1,..., n-1, a relation which defines t_k.

However we would like to have a number n of transition points t_k much less than the size M of a Monte-Carlo sample and still be statistically equivalent to it. Let us assume that our simulation sample has M_k transitions grouped at $t=t_k$ with $\sum_{k=1}^{n} M_k = M$. Then $F^a(t)$ is still of the same type if $M_k/M = 1/n$, for all k. Now $\underset{t}{Max}|F-F^a| = \dfrac{1}{2n}$ must be compared to $1.36/\sqrt{M}$ (and not to $1.36/\sqrt{n}$) and in order that our event simulation be statistically equivalent to a Monte-Carlo sample of size M, the number n of transitions should be proportional to $\sqrt{M}$. This is indeed a very strong requirement.

2.4 Marginal Distributions

We have showed in § 2.2 and 2.3 how two important classes of models derive as particular cases of the general model eq. (4). Further models are characterized by flows in Kirchoff networks (thermal, hydraulic or electric) [24] in which case eq. (1) corresponds to the transient behaviour of the corresponding current, each component x_k being a current in a branch k.

We must emphasize that if we use eq. (4) or the most general formulation, within a semi-markovian approach, there is some allowed flexibility in the modelling of the dynamics eq. (1) which should have an accuracy not exceeding the accuracy expected from an event tree discretization or from a Monte-Carlo.

Let us note, to conclude this chapter, that secondary models can be obtained from eq. (2) by integration or summation. For instance, if we integrate over all $\bar{x}$ variables except x_k with

$$\int \pi(\bar{x},i,t) \prod_{l \neq k} dx_l \equiv \pi(x_k,i,t) \tag{22}$$

$$\frac{\partial \pi(x_k,i,t)}{\partial t} + \frac{\partial}{\partial x_k} \langle f_{i,k}(x_k) \rangle \pi(x_k,i,t)) + \langle \lambda_{i,k}(x_k) \rangle \pi(x_k,i,t)$$

$$- \sum_{j \neq i} \langle p_k(j \rightarrow i|x_k) \rangle \pi(x_k,i,t) = 0 \tag{23}$$

where

$$\langle f_{i,k}(x_k) \rangle \triangleq \int f_{i,k}(\bar{x}) \pi(\bar{x},i,t) \prod_{l \neq k} dx_l / \pi(x_k,i,t)$$

$$\langle p_k(j \rightarrow i|x_k) \rangle \triangleq \int p(j \rightarrow i|\bar{x}) \pi(\bar{x},i,t) \prod_{l \neq k} dx_l / \pi(x_k,i,t) \tag{24}$$

Of course the marginal distribution $\pi(x_k,i,t)$ must be obtained iteratively since functions $\langle f_{i,k} \rangle$ and
$\langle p_k(j \rightarrow i|x_k) \rangle$ are defined implicitly. Finally, we must synthesize the distribution π from its marginal distributions [25,26] a problem which has no unique solution.

3 Methods of Solution and Associated Numerical Problems

3.1 The Curse of Dimensionality

(a)   All methods which truly take into account the dynamics of the system have to integrate the process dynamics eq. (1). Since by sheer necessity the process dynamics will be of moderate complexity (say $N=10$ to 50) the integration of the ODE problem presents no problem. No accuracy beyond what is obtained by Monte-Carlo is needed, the essential choice being one of minimal number of arithmetic operations.

(b)   We would like first to clarify the meaning of discretization, by making a distinction between models and algorithms. A discretization may be associated with (and in fact define) a model like time discretization (in fact branching time) for DYLAM-DETAM or process variables discretization [17].

We have a second level of time discretization which is associated to the solution of eq. (1) in the case of an ODE solver, or of a markovian problem, but in this case the Δt interval can be under control during the calculation and in principle adapted to guarantee a degree of accuracy.

(c)   In the case of event simulation the problem is in principle "solved" with the knowledge of the $\bar{g}_i(t,\bar{x})$ a feature common to all methods. This involves in all cases an ODE solver (a fourth order Runge-Kutta method was used in [2]) which should be robust and simple. If we discretize phase space R^N by p^N blocks, we obtain a markovian problem with $p^N m^n$ states. Direct or even iterative methods are out of question unless drastic methods of simplification are used. Numerical analysts solve time dependent problems in 3 dimensional space with present day computers, using finite difference or nodal or finite elements methods but are skeptical of 4 or 5 dimensions because of the exponential growth of the unknowns. This is precisely the stage where Monte-Carlo method stands without rival. Only small values of N can be hopefully solved by discretization methods. The same curse of dimensionality is attached to pure markovian reliability problems with realistic number of components of many dozens at least and suggest techniques like merging or aggregation.

(d)   The numerical work of Monte-Carlo is linear in the number of games (histories). Discretization of the ODE $\dfrac{d\bar{x}}{dt}=\bar{f}_i(\bar{x})$ yields kN (k=number of intervals) unknowns to be computed and stored for one-step methods, and on the whole, Monte-Carlo is only mildly sensitive to the complexity of the problem. In particular no storage of $\hat{p}(j{\rightarrow}i)$ is necessary. The drawback is of course the number of histories which can be very high if we need very small probabilities of crucial events. However the same difficulty occurs for simulation approaches.

Monte-Carlo has a way to circumvent the difficulty by using modified games i.e. modifying the kernel of eq. (4), but at the same time giving a weight to the trajectory in $(\bar{x},i,t)$ which preserves the expected value of the score. Much contemporary research in Monte-Carlo is devoted to the study of modified games with reduced variance of the score and therefore increased efficiency. Recent work in the field of reliability shows conclusively the activeness of Monte-Carlo for markovian problems [27-31] and for complete phase space simulation [2]. Moreover distribution of transition rates are easily taken into account by Monte-Carlo techniques.

The potential of modified games and in particular the "zero-variance" games has been hardly explored. A Monte-Carlo method is at its best when it is tailored for a specific aim (like evaluation of the probability of crossing a specific safety boundary), in fact for the evaluation of an integral like $\sum\limits_{i} \int \pi\,a(\bar{x},i,t)d\bar{x}$ where $a(\bar{x},i,t)$ is any score relevant to a safety problem.

A typical possibility is the use of the adjoint problem [10,32] instead of the direct problem : trajectories start from the safety boundary after time reversal and transposing the matrix of element $\hat{p}$, and are followed until we reach the domain where the accident is likely to start.

Since solving an integral equation like eq. (4) (and other methods were shown to be essentially approximations of eq. (4)) amounts to doing quadratures. We may remember that if a standard one dimensional quadrature rule has an error of order n^{-m} where n is the number of points, its error is of order $n^{-m/N}$ for a domain of N dimensions although Monte-Carlo is still of order $n^{-1/2}$ for all N [33]. The break-even point is therefore N=2m, for instance N=4 for a trapezoidal rule. Of course we can improve that rule but the same is true for Monte-Carlo if we use stratified sampling [10,34].

Comparing dynamic event tree methods and simulation methods like Monte-Carlo is somewhat like a tale of two travellers. Each will travel randomly and they need maps. The first one buys every map available and if he wants accuracy he will need a van to carry them. The second one stops to buy the current map wherever he needs one and throws it away when he leaves the region. As soon as accurate navigation is needed the second will be best off. Although he may have to buy many times the same one, on the whole the fraction of the country that the travellers visit is very small and most maps are not needed.

It is our personal conclusion that once we leave the domain of small N (N > 4?), Monte-Carlo has no real competitor. Are other methods of solution useless? By no means. In fact we must remember a rule of Monte-Carlo which says that every partial information on the solution, however crude it is, may be used to define a modified game in such a way that our Monte-Carlo budget is used exclusively to obtain the missing information.

Therefore many approximate numerical methods existing or to be developed to solve markovian problems can be used to obtain approximate solutions, sometimes very crude, that in turn will determine the weights in importance sampling, or in zero variance schemes.

3.2 Synthesis

Figure 1 is an attempt to synthesize the mathematical aspects of the methods which explicitly take time into account, as they appear today. We do not believe that the distinction between analytical methodologies on one side and the simulation methodologies on the other, is as clearcut as suggested in [7]. We prefer another method of classification using models on one side and algorithms on the other.

The common core to all methods which are reasonably complete contains the following aspects :

- the description of the physical process and its dynamics;
- the relevant parts of the hardware, especially those liable to fail, to change

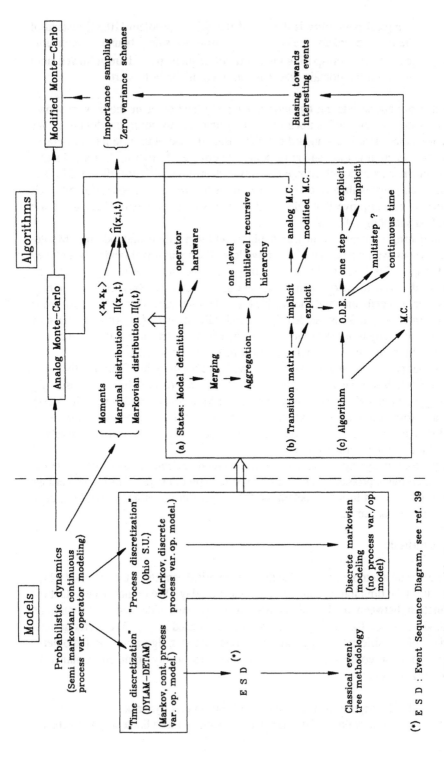

Fig.1. Structure of models and algorithms

(*) E S D : Event Sequence Diagram, see ref. 39

state under automatic control or following a human intervention;
- a modelling of the failure rates in relation with the physical process
  ($p(j \to i)$ in our notation);
- a model of the operator.

The first two are fairly straightforward; the third may involve a lot of information gathering but poses no problem of principle; the last is by far the most controversial since it involves necessarily a description of the operator's "states" as well as transition rules between these states.

The (semi) markovian assumption is not unduly restrictive because it is always possible to include history-dependent rules by the introduction of supplementary states. This step may be costly when explicit transition matrices are required but not if implicitly formulated rules are required for simulation. All methods are in fact analytical in the sense that they can be formulated from a Chapman-Kolmogorov point of view. Different levels of modelization correspond to different types of discretization. These discretizations are inherent to the models and are distinct from the discretization of the numerical algorithms that will be used to solve them.

Algorithms are not unique : for instance a classical markovian problem with discrete states is a model which can be implemented by an algorithm like the solution of a system of ordinary differential equations or by a Monte-Carlo algorithm. Therefore the dichotomy between analytical and simulation methods may not be always relevant. On the other hand a natural dichotomy appears between explicitly and implicitly formulated problems.

The markovian approach in its explicit form is numerically self-defeating : N components each one in m states yield m^N possible states of which a very small part (the more so that N is great) is useful. Conceptually attractive, it is somewhat akin to the use of phase-space in statistical mechanics where N is so large that it is sufficient to study the most probable state. The markovian approach is necessary when components of a system are not independent, i.e. their failure rates are dependent upon the state of other components. In principle the rate of failure of a single component may take m^{N-1} different values, but in practice the rules of dependence are very simple and certainly very few compared to m^N. Therefore there is a good deal of wastage in defining explicitly states and transition probabilities.

Simulation in the sense that a random realization of the transients is obtained[1] needs only to store the actual state of the system and the operational rules. Quoting [7] "it presents a more object oriented view of the system to be modeled and leads itself to the creation of computer models whose objects.... have a one-to-

---

[1] "Simulation" methods like DYLAM-DETAM use probabilistic arguments to enumerate the most probable branches of event trees but do not contain any random element in the computational process itself.

one correspondence with actual objects in the system being simulated". Objections to simulation concern the need for postprocessing and difficulties in determining the importance of parameters. The first one is real only if the postprocessing computer time is a sizeable amount of the simulation computer time and the second is not necessarily warranted if one uses the adjoint formulation [32] which gives precisely the "importance" of a state $(\bar{x},i)$ in producing a damage.

Finally the whole process can be analyzed according to the stages defined below.

1. The first stage (and most important) is the definition of the model as alluded above.
2. The second is the choice of the objective(s) of the analysis : probability of crossing a safety boundary, importance of a component (or a control law) in the occurrence of a transient, etc. General purposes methods are generally less efficient than objective-tailored methods.
3. The next is the choice of the Chapman-Kolmogorov equation adapted to the objectives of the problem (differential vs integral, direct vs adjoint).
4. The following (which may be bypassed for analog Monte-Carlo) introduces the basic discretizations that allow the problem to be solved numerically.
5. The next involves the actual definition of the numerical algorithms which involves (for the case of ODE for instance) nearly always (after due linearization through a Newton process) the solution of a real or virtual linear system of algebraic equations, the latter in the case of Monte-Carlo
6. The final is the realization of the computation itself either by Monte-Carlo or by a suitable iterative method (or direct methods if the method of solution is time-explicit like Euler's, if the stiffness of the problems allows).

Let us remind finally that the use of modified Monte-Carlo by variance reducing schemes may well involve both explicit and implicit approaches. Moreover Monte-Carlo practitioners remind us that reduction of variance techniques are indeed powerful but sometimes tricky and in need of fine tuning.

References

1. Devooght J., Smidts C.: Probabilistic reactor dynamics. I. The theory of continuous event trees. Nucl. Sci. Eng. 111, 229-240 (1992)
2. Smidts C., Devooght J.: Probabilistic reactor dynamics. II. A Monte-Carlo study of a fast reactor transient. Nucl. Sci. Eng. 111, 241-256 (1992)
3. Gardiner C.W.: Handbook of stochastic methods. 2nd ed. Berlin: Springer 1985
4. Ingman D., Reznik L.: Dynamic character of failure state in damage accumulation processes. Nucl. Sci. Eng. 107, 284 (1991)
5. Evaluation of severe accident risks : quantification of major input parameters. Nureg/CR-4551. Sand 86-1309. Vol. 2. Rev. 1, Part 2. U.S. Nuclear Regulatory Commission (1991)

6.   Birolini A.: On the use of stochastic processes in modeling reliability models. Lect. Notes in Economics and Mathematical Systems 252. Springer 1985

7.   Siu N.: Risk assessment for dynamic system : an overview. Submitted to Reliab. Eng. and Systems Safety, April 1992

8.   Matkowski B.J., Schuss Z.: The exit problem for randomly perturbed dynamical systems. SIAM J. Appl. Math. 33, 230 (1970)

9.   Schuss Z.: Theory and applications of stochastic differential equations. New York: John Wiley and Sons 1980

10.   Lux I., Koblinger L.: Monte Carlo particle transport methods : neutron and photon calculation. CRC Press 1991

11.   Pages A., Gondran M.: Fiabilité des systèmes. Collection DER EDF, Vol. 39. Eyrolles 1980

12.   Bharucha-Reid A.T.: Elements of the theory of Markov processes and their application. New York: McGraw-Hill 1960

13.   Aldemir T.: Computer assisted Markov failure modeling of process control systems. IEEE Transactions on reliability R-36, 133-144 (1987)

14.   Belhadj M., Hassan M., Aldemir T.: The sensitivity of process control system interval reliability to process dynamics. A case study. Proceedings PS AM Mtg., Los Angeles 1991

15.   Aldemir T.: Utilization of the cell to cell mapping technique to construct Markov failure models for process control systems. Proceedings PS AM Mtg., Los Angeles 1991

16.   Aldemir T.: Quantifying setpoint drift effects in the failure analysis of process control systems. Reliab. Eng. Sys. Saf. 24, 33 (1989)

17.   Hassan M., Aldemir T.: A data base oriented dynamic methodology. Reliab. Eng. Syst. Safety 27, 275-322 (1990)

18.   Hsu C.S.: Cell-to-cell mapping. Applied Mathematical Sciences 64. Springer 1987

19.   Amendola A.: Accident sequence dynamic simulation versus event trees. Reliab. Eng. Syst. Safety 22, 3 (1988)

20.   Amendola A., Reina G.: Event sequences and consequence spectrum : a methodology for probabilistic transient analysis. Nucl. Sci. Eng. 77, 297 (1981)

21.   Amendola A., Reina G.: Dylam-1, A software package for event sequence and consequence spectrum methodology. Eur 9224 En. Commission of European Communities (1984)

22.   Cacciabue P.C., Carpignano A., Vivalda C.: Expanding the scope of Dylam methodology to study the dynamic reliability of complex systems : the case of chemical and volume control in nuclear power plants. Reliab. Eng. Sys. Saf. 36, 127 (1992)

23.   Siu N., Acosta C.: Dynamic event tree analysis. An application to SGTR. Proceedings Int. Conf. Probabilistic Safety Assessment and Management, Beverly Hills 1991, p. 413. Elsevier Science Publishers 1991

24.   Arien B., Lamy D., Devooght J., Smidts C.: Reliability analysis of large systems by the markovian technique, development of the CAMERA software. Colloque international sur l'utilisation de l'évaluation probabiliste. AIEA. Vienna 1991

25.   Labeau P.E., Devooght J.: Synthesis of distribution for probabilistic reactor dynamics. To be presented at M&C/SNA '93 Conference, Karlsruhe 1993

26. Devooght J., Smidts C.: Probabilistic reactor dynamics : computational models. Trans. Amer. Nucl. Soc. 64, 289-290 (1991)

27. Dubi A. et al.: Analysis of non markovian systems by a Monte-Carlo method. Annals of Nucl. En. 18, 125-130 (1991)

28. Yen-Fu Wu, Lewins J.D.: System reliability perturbation studies by a Monte-Carlo method. Annals of Nucl. En. 18, 141-146 (1991)

29. Marseguerra M., Zio E.: Non linear Monte-Carlo reliability analysis with biasing towards top events. To be published in Reliab. Eng. Syst. Safety.

30. Zhuguo T., Lewis E.E.: Component dependency models in Markov Monte Carlo simulation. Reliab. Eng. 13, 45 (1985)

31. Lewis E.E., Zhuguo T.: Monte Carlo Reliability modeling by inhomogeneous Markov processes. Reliab. Eng. 16, 277 (1986)

32. Buslik A.: Monte-Carlo methods for the reliability analysis of Markov systems. Trans. Amer. Nucl. Soc.

33. James F.: Monte-Carlo theory and practice. Reports on Progress in Physics 43, 1145-1189 (1980)

34. Press W., Farrar G.: Recursive stratified sampling for multidimensional Monte-Carlo integration. Comp. in Physics 4(2), 190-195 (1990)

35. Martz H.F., Waller R.A.: Bayesian reliability analysis. Wiley 1982

36. Papazoglou I., Gyftopoulos E.P.: Markovian reliability analysis under uncertainty with an application on the shutdown system of the Clinch River Breeder reactor. Nucl. Sci. Eng. 73, 1-18 (1980)

37. Smidts C.: Probabilistic reactor dynamics. IV. An example of man/machine interaction. Nucl. Sci. Eng. 112, 114-126 (1992)

38. Devooght J., Smidts C.: Probabilistic reactor dynamics. III. A framework for time dependent interaction between operator and reactor during a transient involving human error. Nucl. Sc. Eng. 112, 101-113 (1992)

39. Apostolakis G., Chu T.L.: Time dependent accident sequences including human actions. Nucl. Technology 64, 115-126 (1984)

Evaluating Performance and Reliability of Automatically Reconfigurable Aerospace Systems Using Markov Modeling Techniques

Bruce K. Walker

Department of Aerospace Engineering & Engineering Mechanics
University of Cincinnati, OH 45221-0343, U.S.A

Abstract. In this paper, we consider the use of Markov modeling techniques for the evaluation of reliability and performance of highly reliable aerospace systems that involve automatic failure detection and system reconfiguration capability. The presence of these automated functions makes the system configuration dynamic, and necessitates the use of dynamic analysis techniques to analyze the system reliability. The paper surveys some of the Markov modeling techniques that can be used for this purpose and the types of situations to which they can be applied. The paper also briefly describes a method for using a Markov model to evaluate quantities other than the overall system reliability that describe the performance of the system. The values of these quantities can depend upon the time history of the system configuration, therefore they can only be evaluated based on a dynamic analysis technique.

1 Introduction

The reliability requirements placed on current aerospace vehicles and their subsystems often exceed greatly the reliability that can be achieved by single string, nonredundant components, even when the highest quality components with state-of-the-art reliability are used. In order to satisfy these very high reliability requirements, many aerospace vehicles and systems include substantial component redundancy and provide "fault tolerant" operation despite the occurrence during a mission of component faults.

The control system for aerospace vehicles and systems is often a vital element in acceptably operating these systems, particularly when the system is unstable in the absence of feedback control, which is true for such vehicles as the F-16, the X-29, and essentially all rotary-wing aircraft. In these applications, the control system is safety-critical, and therefore the reliability requirements on the control system can be as high or higher than the vehicle structure. Because the individual components comprising these systems rarely achieve such high reliability levels, the system reliability must be achieved despite the likely occurrence of component failures, i.e., the system must be "fault tolerant". In order to achieve the desired level of fault tolerance, many current aerospace control systems include redundant components. In particular, redundant control surfaces, redundant actuation elements, redundant sensors, and redundant power supply elements are frequently included.

In addition to redundant components, a fault tolerant aerospace control system must also have a means for detecting and identifying component failures and for reconfiguring the control system to compensate for the failures that have been identified. Unlike many process control and power generation systems, a control system component failure in these systems can result in irrecoverable loss of control in a few seconds or even in a fraction of a second. Human pilots cannot respond quickly enough to save the vehicle in these settings, and therefore the control system must include software that automatically implements redundancy management (RM) logic to perform the failure diagnosis and control reconfiguration tasks. This logic must operate in real time, and therefore the configuration of the system is dynamic with transitions possible in very short time intervals. Furthermore, the RM logic must operate on measurement data from the sensors that are corrupted by noise and are also affected by disturbances that affect the system that is being controlled. As a result, the RM logic is subject to random errors.

Any method for assessing the reliability or performance of a system that includes automatic RM logic of the type just described must account for the dynamic nature of the system configuration and the possibility of random RM logic errors. One framework for modeling the system behavior that lends itself to such analyses is Markov modeling, which can be shown to have significant advantages relative to combinatorial techniques [1,2].

A Markov model describing the dynamic behavior of the configuration of a fault tolerant control system consists essentially of a finite set of states associated with the various possible system configurations and conditions and the associated transition probabilities [3,4]. We assume here that the RM logic is implemented in discrete time and that therefore the behavior model will be described in discrete time. The transition probabilities are constructed from the probabilistic description of component failure events and of RM logic events. This is possible provided the random failure events and RM logic behavior are memoryless. Since failure event behavior is usually modeled by the exponential distribution, the failure behavior is typically memoryless. The RM logic behavior will also be memoryless if the RM logic decisions are based only on instantaneous data with random components that are uncorrelated with past data or with the past history of the system configuration. For RM logic satisfying these conditions, the Markov model can be constructed and the system reliability can be evaluated by solving for the transient solution of the model.

In many aerospace systems, the statistical tests used in the RM logic include data memory to reduce the error probabilities associated with the RM tests. As a result, the random behavior of the system cannot be described by a simple Markov model. For many of these systems, it is possible to construct a semi-Markov model for the system behavior [5,6], in which case the transition probabilities are replaced by transition time probability mass functions (pmf's), where we assume again that the behavior model is to be described in discrete time.

Once a model is constructed, the problem of evaluating the system reliability becomes one of evaluating the transient state distribution probabilities for this Markov or semi-Markov behavior model. The size of the model, in terms of the number of states, is often very large, however, even for relatively simple systems subject to only a few random behavior elements. Finding the solution by numerical means then becomes problematic. We will discuss in this paper an approximation method that reduces the need for numerical computations.

In many aerospace applications, the performance of the control system is just as important as its reliability. For example, the value of a high precision automatic navigation system is judged based upon its accuracy in determining the position and velocity of the vehicle in which it is installed. This accuracy must be achieved despite the presence of noise and disturbances in the signals that are processed. If the system is also designed with redundant components and RM logic in order to achieve fault-tolerance, then the accuracy must be assessed in a way that includes the effects of random failures and RM logic errors. In many cases, it is possible to define a performance metric that can be related to the history of the dynamic system configuration. In the navigation system example, the expected rms position estimation error is an example of such a metric.

The Markov modeling techniques used to evaluate the system reliability can be used to evaluate the statistical properties of configuration-dependent performance metrics. However, this usually requires the exhaustive evaluation of the probability of every possible trajectory for the system configuration. Again, the size of the model, which can be large even for simple system designs, leads to a numerical evaluation problem of monumental size, thereby motivating techniques for approximating the solutions, one of which we discuss here.

In this paper, we summarize some results on the use of Markov modeling techniques to evaluate the reliability and performance of fault tolerant aerospace control systems that include automatic RM logic subject to random errors.

We begin with a discussion of the extension of Markov modeling techniques to situations where the RM tests are not memoryless, and therefore the Markov property is violated by these tests. The key to this extension is the use of semi-Markov process theory, which we briefly explain. We present there a limit theorem that we developed for approximating the transient solutions to these models in a computationally practical way. We illustrate the use of the theorem with two simple example systems. Then, we present the concept of performance for a fault tolerant system and we present a method for evaluating the probability mass function (pmf) of the value of a performance metric using a new transform technique. The performance pmf is then generated for a realistic aerospace system example.

2 Reliability Analysis by Semi-Markov Modeling Methods

In order to reduce the effects of measurement noise on the failure detection and identification (FDI) process in aerospace control systems, RM logic designers often make use of sophisticated filter designs to generate the data used to make the FDI decisions and of powerful statistical tests to make the logical decisions. Examples of specially designed data filtering methods for FDI include the failure detection filter [7,8,9], the generalized parity equation method [10,11], and recent efforts using the unknown input filter approach [12,13]. Sophisticated statistical testing approaches that have been formulated for FDI use include the sequential probability ratio test [14], the generalized likelihood ratio (GLR) test [15,16], the one-sided sequential reset detection test [17], and the Shiryaev test [18].

Approaches to FDI that use these techniques all share the property that the probabilistic likelihood of any particular outcome of an FDI test at a specified time sample is not statistically uncorrelated to the data that were used in the previous FDI tests or the outcomes of the previous FDI tests. This violates the Markov property that is exploited when Markov models are constructed for the system behavior, and therefore renders Markov modeling inapplicable to systems employing these tests.

For many of these strategies, however, there are conditions under which the tests are "reset" or "renewed," usually involving logic related to the RM logic. For example, two-sided sequential FDI tests can result in a "no failure" decision. When this decision is reached, most RM logic designs would call for the test to be reinitiated, i.e. the test would be restarted. Even for FDI test procedures that do not reach "no failure" decisions, the RM logic often involves occasional resetting or the use of a limited window of past data to avoid biasing the tests strongly away from detection decisions in order to avoid long detection delays when a failure does occur.

For FDI strategies that involve occasional resetting, the outcomes of the tests following the reset are statistically uncorrelated to the outcomes of the tests preceding the reset provided the data used by the tests following the reset include random components that are uncorrelated with the data preceding the reset. In these cases, a model for the system behavior can be constructed using semi-Markov process (or Markov renewal process) theory [19, 20].

A finite-state, discrete time semi-Markov process is similar to a finite-state, discrete time Markov process in that a finite set of states must be defined and the transition behavior among the states must be defined, where it is assumed that the statistical behavior of any particular transition depends only upon the state exited by the transition and the state to which the transition leads and not on the past history of transitions that led to the original occupancy of the state exited. The difference between the two processes is that a semi-Markov process allows the distribution of the time required to make the transition to have an arbitrary distribution, while Markov models require the transition time distributions to be exponential (and therefore memoryless in the sense that the remaining time to the transition is independent of how much time has elapsed since the last transition).

The construction of a semi-Markov model for a fault-tolerant control system is similar to that of constructing a Markov model. The states are defined in the same way to reflect the system configuration and the RM logic option that is being used. The events leading to a transition are also formulated in the same way. The difference is that the probability mass functions (pmf's) of the time required to make each possible transition must now be developed. Since each transition involves some combination of failure events and FDI test outcomes, the pmf of the time required for these joint events must be constructed. Assuming that the pmf of the time required for each FDI test event under the failure conditions implied by each state in the model are known, that the FDI tests are independent of one another, and that the failure time distributions are exponential, the pmf's of the transitions can be constructed. See [5] or [6] for details.

The evaluation of the transient history of the state probability distribution for the process then proceeds by standard semi-Markov process theory. Using the notation of [19], let G(m) be the "core matrix" for the semi-Markov process, where the (i,j) element of G(m) is defined as follows:

$$g_{ij}(m) = \text{Prob\{transition occurs from i to j exactly m time samples after i is entered\}}$$

and let $\pi(n)$ be the row vector representing the state probability distribution at time sample n, as was the case for Markov process models. Assume that the initial state distribution $\pi(0)$ is given. Then:

$$\pi(n) = \pi(0) \, \Phi(n)$$

where $\pi(n)$ is the n-step transition probability matrix, which is given by:

$$\Phi(n) = W^>(n) + \sum_{m=1}^{n} G(m) \, \Phi(n-m)$$

where $W^>(n)$ is given by:

$$W^>(n) = \text{diag}\{ 1 - \sum_{m=1}^{n} \sum_{k=1}^{n} g_{ik}(m) \}$$

which can also be constructed recursively as:

$$W^>(0) = I$$

$$W^>(l) = W^>(l-1) - \text{diag}\{ \sum_{k=1}^{n} g_{ik}(l) \}$$

Notice the presence of a convolution sum in the expression for $\Phi(n)$. Recall from the discussion of Markov models above that the number of states in a model for even a simple fault tolerant system can be relatively large. Now, if a semi-Markov process model must be used to describe the system's random behavior, the computational problems associated with evaluating a model of large dimension are complicated further by the need to evaluate this convolution sum. In cases where the elements of $G(m)$ become vanishingly small for relatively small values of m, this is not a major problem. Unfortunately, for most FDI tests the typical time to decision can be quite long (especially in the absence of a failure), and therefore the elements of $G(m)$ do not become small except for very large values of m. This requires the evaluation of many of the terms in the convolution sum, which can quickly become impractical when the model dimension is large.

Over the past few years, we have generated a number of results regarding the approximate construction of solutions for the transient history of semi-Markov process models of fault-tolerant aerospace system behavior. These approximations use decomposition techniques that break down the model into "pieces" of smaller dimension that can be more easily evaluated. Then the results for the "pieces" are reassembled to approximate the solution for the original process model.

Our key result along these lines is the following theorem, which is stated and proven in [21]:

THEOREM 1 (Limit Theorem for Semi-Markov Chains) Let the set E of states of the semi-Markov chain be expressible as a union of disjoint classes:

$$E = \sum_{k=1}^{N^e} E_k \ k \in M \equiv \{1, 2, ..., N^e\}$$

Let $\tau_{kr}^{(i)}$ be the sojourn of the semi-Markov chain in class E_k when it starts from state $i \in E_k$ and moves to class E_r where $r \neq k$. Suppose the following two conditions hold for the semi-Markov chain E:

1. The elements of the core matrix sequence $\{g_{ij}^\varepsilon(n) \mid i, j \in E\}$ specifying the semi-Markov chain depend as follows on the small parameter ε:

$$g_{ij}^\varepsilon(n) = p_{ij}^\varepsilon h_{ij}^\leq \left(\frac{n}{\varepsilon}\right)$$

where $h_{ij}^\leq(0) = 0$. Furthermore, the p_{ij}^ε can be expanded in a Taylor series about $\varepsilon = 0$, yielding:

$$p_{ij}^{\varepsilon} = \begin{cases} p_{ij}^{(k)} - \varepsilon q_{ij}^{(k)} + 0(\varepsilon) & \text{if } i, j \in E_k \\ \varepsilon q_{ij}^{(k)} + 0(\varepsilon) & \text{if } i \in E_k \text{ and } j \notin E_k \end{cases}$$

where $0(\varepsilon)$ represents terms of order higher than one in ε (i.e., $0(\varepsilon)/\varepsilon$ approaches zero as ε approaches zero).

The embedded Markov chain for $\varepsilon = 0$, (that is, the unperturbed Markov chain) obeys the usual Markov chain properties:

$$\sum_{j \in E_k} p_{ij}^{(k)} = 1; \text{ and } p_{ij}^{(k)} \in [0,1]; \ \forall \ k \in M$$

2. The embedded Markov chains defined by the matrices $\{ p_{ij}^{(k)} \mid i, j \in E_k \forall k \in M\}$ are ergodic (non-ergodic with one and only one unit magnitude eigenvalue), with stationary (or Caesaro limit) probabilities $\{ \pi_i^{(k)} \mid i \in E_k \forall k \in M\}$.

Then

$$\lim_{\varepsilon \to 0} \Pr\{\tau_{kr} \leq t\} = \gamma_{kr} \left\{ 1 - \exp\left[\frac{-\Lambda_k t}{T} \right] \right\},$$

where

$$\gamma_{kr} \equiv \frac{\sum\limits_{i \in E_k} \pi_i^{(k)} q_i^{(kr)}}{\sum\limits_{i \in E_k} \pi_i^{(k)} q_i^{(k)}} ; \quad \Lambda_k \equiv \frac{\sum\limits_{i \in E_k} \pi_i^{(k)} q_i^{(k)}}{\sum\limits_{i \in E_k} \pi_i^{(k)} a_i^{(k)}} .$$

Here:

$$q_i^{(kr)} \equiv \sum_{j \in E_r} q_{ij}^{(k)}, \ q_i^{(k)} \equiv \sum_{j \in E_k} q_{ij}^{(k)} .$$

$$a_i^{(k)} \equiv \sum_{j \in E_k} p_{ij}^{(k)} \bar{\tau}_{ij}, \ \bar{\tau}_{ij} \equiv \sum_{n=0}^{\infty} n \, h_{ij}(n)$$

The interpretation of this theorem is as follows. Suppose that the semi-

Markov model of the system behavior can be split up into classes of states where the transitions within a class are very likely and tend to occur over relatively short time intervals while the transitions to states outside the class are rare. This is essentially the meaning of condition 1 in the theorem, although the condition imposes some additional conditions on the mathematical forms that the transition probabilities and transition time mass functions can take. Suppose in addition that if we neglect the interclass transitions and consider only the transitions within a class (by setting the interclass transition probabilities to zero and adjusting the intraclass transition probabilities to reflect the loss of possible transitions), we find that each of the intraclass embedded Markov processes is either ergodic or nonergodic but with one and only one unit magnitude eigenvalue (i.e. with only one trapping class). The latter condition allows us to find a limiting state probability distribution for the intraclass process. Under these conditions, we can treat the long-term interclass behavior of the model as a Markov chain with class-to-class transition probabilities given by γ_{kr} and the exit time rate for class k given by Λ_k. Here, both γ_{kr} and Λ_k are determined by the limiting distributions that we find for the intraclass processes when the interclass transitions are suppressed, by the relative sizes of the small interclass transition probabilities, and by the mean values of the transition time distributions within each class.

From the long-term transient behavior of the probabilities of occupying each class determined by the limit theorem and from the intraclass limiting distributions, we can construct an approximate solution for the transient behavior of the original semi-Markov model by a disaggregation step as follows:

Let $\hat{\underline{\pi}}^e(t)$ be the vector of estimated <u>class</u> occupation probabilities from the limit theorem. Then:

$$\hat{\underline{\pi}}(t) = [\hat{\pi}_1^e(t)\,\underline{\pi}_1 \mid \hat{\pi}_2^e(t)\,\underline{\pi}_2 \mid \cdots]$$

is the estimated vector of state probabilities where $\underline{\pi}_k$ are the stationary (or Cesaro limit) probability distributions within the class when interclass transitions are eliminated (i.e. when $\varepsilon \to 0$).

This approximation technique has major implications for reducing the computations necessary to evaluate the long-term transient solutions of the large dimension semi-Markov models that represent fault tolerant system behavior. Typically, the likelihood of component failures in a fault tolerant system is very small relative to the likelihood of RM logic events (such as false alarms). Also, the RM logic event behavior tends to be quite fast relative to the typical time required for a component failure event to occur. Therefore, the states of a semi-Markov model for fault tolerant system behavior tend to group into classes according to the failure status of the system. The transitions associated with failure events then become the rare interclass transitions and the RM logic event transitions become the relatively fast and frequent intraclass transitions. Thus, the first condition of the limit theorem is satisfied with the small parameter ε corresponding to the component failure probabilities, provided the mathematical form of the transition probabilities agrees with that of the condition, and this is not usually a problem.

The next step in applying the theorem is to check on the satisfaction of the second condition, which is a condition on the embedded Markov process of the intraclass process when the interclass transitions are suppressed. Since the interclass transitions in a model for fault tolerant system behavior are associated with failures, the suppression of interclass transitions corresponds to neglecting the failure events. Hence, the intraclass process of interest is determined completely by the RM logic behavior with the failure status of the system held fixed. This is a crucial observation because the behavior of the RM logic under fixed failure conditions is often examined and characterized as part of the RM logic design process, and therefore is frequently known before the reliability and performance evaluation process is started.

In automatic fault tolerant control system implementations, the RM logic form at each level of the RM strategy can usually be characterized as one of two types: recoverable logic and irreversible decision logic. Recoverable logic includes all RM strategies whereby a component flagged previously as failed (or suspected failed) can be brought back online. This type of logic is often used when false detection alarms are very likely and the loss of use of a component due to a false alarm is high enough that the investment in the added logic complexity required to further test the flagged component and bring it back online is warranted. Irreversible decision logic characterizes all RM logic strategies where the indication of the failure of a component by the automatic fault diagnosis scheme results in permanent elimination of that component from further use. This type of logic is simpler to implement than recoverable logic, but its use incurs the penalty that false alarms can permanently remove good components from use.

The second condition of the theorem requires that the intraclass transition behavior have an embedded Markov chain that is either ergodic or nonergodic but with exactly one unit magnitude eigenvalue. When recoverable logic is used at some level of the RM logic, the resulting semi-Markov model of the system behavior generally leads to intraclass behavior with an ergodic embedded Markov process, and the condition of the theorem is satisfied. On the other hand, the use of irreversible decision logic as part of the RM strategy usually results in the presence of trapping states within each class when the interclass transitions are suppressed. Therefore, the intraclass behavior is not ergodic. However, the weakened condition regarding the presence of no more than one unit magnitude eigenvalue for the embedded Markov process is frequently satisfied because each class often contains only one trapping state or only one periodic trapping set. Thus, for most models of fault tolerant system behavior, at least the weaker condition on the embedded Markov process associated with the intraclass transitions is satisfied.

The remaining steps in the application of the theorem involve calculation of the quantities needed to find γ_{kr} and Λ_k. Except for the stationary (or Caesaro limit) state probability distributions for the embedded Markov process associated with the intraclass behavior, the quantities required for these calculations come directly from the original model after its transition probabilities are decomposed into the ε-dependent part and the non-ε-dependent part. For each class of states where the intraclass transition behavior is ergodic, the stationary distribution can be found by standard techniques from Markov process theory [3,4,20]. For classes where the Caesaro limit distribution must be found, the limitation that only a single unit magnitude eigenvalue can be present for the embedded process allows us to solve for a single eigenvector to find the desired distribution.

To illustrate the use of the limit theorem, we repeat here results that originally appeared in [21]. The system to be considered is a very simple single component system with continuous monitoring of the component for failure by a hypothetical sequential monitoring test. The time to a false detection decision for the monitoring test is assumed to have a hypergeometric distribution with two degrees of freedom, and such decisions are assumed to occur with probability P_{fa}. The time to a correct decision that no failure is present is also assumed to obey a different hypergeometric distribution with two degrees of freedom, and occurs with probability $1-P_{fa}$. The test's decision properties when the component has failed are not an issue in this case because failure of the component results in loss of system operation regardless of the monitoring test outcome.

Admittedly, this example system is extremely simple and not reflective of the complexity inherent to real fault tolerant systems. However, it was selected because its simplicity allows us to solve analytically for the transient history of the state probability distribution of the resulting semi-Markov model. This then provides exact answers to which the approximate answers generated by the use of the limit theorem can be compared.

Figure 1 shows the schematic of the semi-Markov model for the behavior of this system. The transitions represented by solid lines in the Figure are transitions due to decisions by the sequential monitoring test. Those represented by dashed lines are transitions that involve failure of the component. The model has been decomposed into two classes, as indicated in the Figure. Note that interclass transitions are associated with component failures.

Figure 2 (taken from [21]) shows the results of a numerical case for this example system. For the numerical results, the MTBF of the component was assumed to be 4000 sec. while the hypergeometric distributions for the times to decision of the monitoring test were determined such that the mean time to a false alarm decision was 16 samples and the mean time to a correct no-failure decision was 36 samples. By selecting an intersample time of 0.2 sec., this yields 3.2 sec. and 7.2 sec., respectively, for these two mean times to decision, which clearly is much smaller that the component MTBF. P_{fa} was selected as .05.

111

Figure 1. Semi-Markov transition diagram for SCMS-I

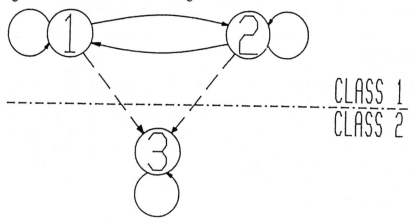

Figure 2. State probability time histories for SCMS-I. (a) Analytical and approximate
solutions, (b) Relative error

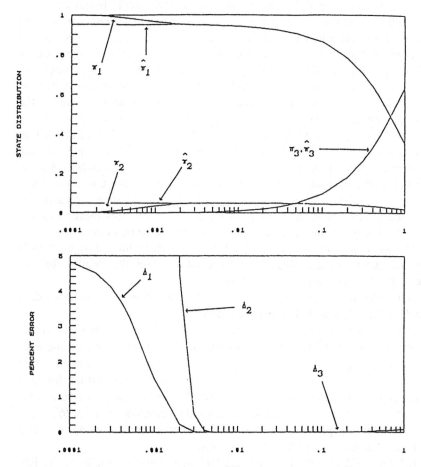

NORMALIZED TIME (T/MTBF)

The top plot in Figure 2 shows the analytically computed exact state probability histories (π_1, π_2, π_3) and the approximate state probability histories ($\hat{\pi}_1$, $\hat{\pi}_2$, $\hat{\pi}_3$) obtained by using the limit theorem to compute the approximate class probabilities and then forming the approximate state probabilities by the disaggregation step. The results are plotted against normalized time on a log scale so that the approach of the approximate results to the exact results can be seen clearly. The bottom plot shows the percentage error of each approximation. Notice that the error is initially significant, but as the time elapsed approaches the order of magnitude of the MTBF, the error quickly converges to near zero. Values of time beyond the MTBF are not shown because a fault tolerant aerospace system would typically not be operated for a duration exceeding the MTBFs of its components without some type of scheduled preventative maintenance.

Occasionally, systems with automatic fault tolerance will yield a behavior model that violates one of the conditions of the limit theorem. In this case, use of the limit theorem to predict performance will lead to inaccurate results. One such situation that arises frequently in aerospace practice is of a system that uses a sequential fault diagnosis test or RM logic specifically designed to produce very long holding times for false alarm decisions, even though the holding times for correct detection and for correct no-failure decisions may be short. In this case, the assumption that the holding time mass functions for the transitions are short relative to the mean time between failure events is violated. This can have a profound effect on the results when the class of states that includes a state for which this test would be in effect also has a single trapping set (when interclass transitions are suppressed) that does not include this state. In this case, the limiting state probability distribution within this class would assign a value of zero for the probability of occupying the state (or states) for which the long holding time test is in effect. In effect, this means that the approximation treats this state as a fast transient state when in fact it is not. If this state has nonzero transition probabilities associated with exiting its class to states in other classes, this can considerably skew the results for the probabilities of occupying the other classes.

As an example of this situation, consider the single component, dual redundant (SCDR) system first considered in [21].

The SCDR system consists of two identical components, designated the primary component and the backup component. At least one of the components must be operating normally for the system to be operational. We assume that the system can operate temporarily using a failed component, but not for long periods in time. In particular, we assume that the use of a failed component for the length of time required to reach the first failure isolation decision following the failure is acceptable, but that any further operation using a failed component results in catastrophic system loss.

In [24], the SCDR system is analyzed for some simple examples of the RM logic. Here, we consider the SCDR with the more complex, two-stage RM logic used in [22,25]. Two sequential tests are used to monitor the difference between the outputs of the components for failures, one test looking for positive bias and the other for negative bias. Each test is a one-sided test, i.e. it can reach only the decision that the respective bias is present, not that it is not present. Once either of the detection tests is triggered, two independent, two-sided SPRTs are initiated for isolation, one for each component, driven by that component's output and the predicted output from an external source. If both isolation tests reach failure decisions, they are reset and reinitiated. If both reach no-failure decisions, the detection alarm is rejected and the detection tests are reinitiated. If one isolation test reaches a failure decision and the other a no-failure decision, the appropriate component is isolated. It is assumed that all of the sequential tests have times to decision that are hyper-geometrically distributed with two degrees of freedom.

The RM logic selects the primary component for use until one of the components is isolated as failed. At that point, the logic calls for use of the component that has not been isolated as failed, and failure detection processing is discontinued. Therefore, this logic is an example of irreversible decision logic. Such logic is justifiable here because the assumption that the system cannot maintain satisfactory operation through more than one failure diagnosis cycle means that even if failure detection processing were continued following the incorrect isolation of the failed component, correct isolation of the component on the next decision cycle would occur too late to rescue the system. The system can be lost if any of the following events or combinations of events take place: 1) Both components fail, 2) One component fails but then the other is isolated, 3) One component is falsely isolated as failed and then the remaining component fails. Taking into account the possible FDI test events and the failure events, we produce a 9-state semi-Markov model for the behavior of this system, which is shown in Figure 3.

States 1, 2, and 3 of the model involve no component failures, states 4, 5, 6, 7, and 8 involve one component failure, and state 9 involves two component failures. The RM status corresponding to each state is as follows:

State   RM status

1 No alarms, detection tests running
2 Detection alarm (false), isolation tests running
3 One instrument isolated, RM disabled
4 Correct detection alarm, isolation tests running
5 Incorrect detection alarm, isolation tests running
6 No alarms, detection tests running
7 Failed component correctly isolated, RM disabled
8 Wrong component isolated (system loss), RM disabled
9 (RM status irrelevant, system loss)

Figure 3. Nine-state transition diagram for SCDR system

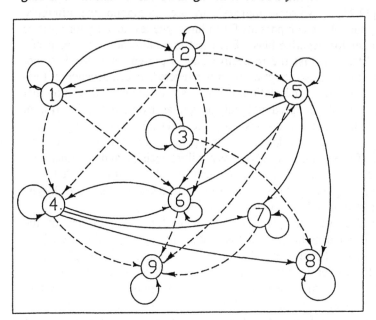

Because the component failure events occur with small probability, the grouping of the states into classes corresponds to the grouping above by failure status, i.e. Class 1 comprises states 1, 2, and 3, Class 2 comprises states 4, 5, 6, 7, and 8, and Class 3 comprises state 9. The solid lines in Figure 3 represent fast transitions (not dependent on the failure rate) while the dashed lines represent slow transitions that involve component
failure events.

When the slow, interclass transitions are neglected, Class 2 for this model includes two trapping states (7 and 8). As a result, Theorem 1 given above is not applicable to this model because the condition that the intraclass embedded Markov process for each class have no more than one eigenvalue of unity will be violated. Instead, the extension of the limit theorem to cases involving multiple trapping classes given in [25] must be used. Like the results in Theorem 1, the extended result involves approximating the interclass behavior of the model by a Markov process that evolves in a slow time scale and then disaggregating the approximate class probabilities into approximate state probabilities using a static approximation to the state probability distribution within each class (see [25] for details on determining this static approximation).

Notice that within Class 1, state 3 is a trapping state and within Class 2, states 7 and 8 are trapping states after interclass (slow) transitions are neglected. As noted above, this means that the static approximations to the intraclass state probability distributions used in the modified limit theorem will assign zero probability to all of the other, transient states (1, 2, 4, 5, and 6). But the limit theorems use this static approximation in deriving the approximate interclass behavior (see the expressions for γ_{kr} and Λ_k above and in [25]). Therefore, if holding times of significant lengths relative to the interclass transition times occur in some of these transient states, the results given by the approximation can be significantly in error. For fault tolerant systems with long holding times to false alarm decisions (which is usually a desirable property), this can result in incorrect approximations to the reliability.

To mitigate this situation, we suggest in [22] a modified approximation approach. The modification basically involves replacing the constant interclass transition rate Λ_k exiting Class k derived above with a time-varying rate $\Lambda_k(t)$, which is determined by summing the exit rates from each of the states within Class k weighted by the transient state probability distribution within Class k (with interclass transitions neglected) at time t. The initial condition for this transient solution is determined either by the given initial condition or by assuming an initial distribution within the class that is consistent with the entry rates from the states that generate transitions that enter the class. The latter method is contingent on the assumption that the interclass transitions are unidirectional in the sense that all transitions out of one class lead only to classes from which transitions back to the originating class are not possible. In fault tolerant system models, this is often true, especially when irreversible RM logic is used, because the tendency of the system is to become more degraded as time elapses.

To avoid computational difficulties with the nonhomogeneous Markov model that approximately describes the interclass behavior in this case, we assume that $\Lambda_k(t)$ is constant over intervals that we call the "time step". We then need only solve the transient intraclass models (with interclass transitions neglected) for their solutions at integer multiples of the time step. The complete procedure is detailed in [25].

As a numerical example, we consider the SCDR described above. We assume the component failure rate is such that the probability of a component failure during one time sample is $5 \times 10(-8)$ for one case and $5 \times 10(-5)$ for another case. The single-sided tests for detection are assumed to terminate with a false detection decision within 5 samples with probability .047 with a hypergeometric distribution of the time required for such termination that has a mean of approximately 3 samples. The low probability of early termination means that the average occupancy time in state 1 in the absence of a failure is typically hundreds of samples. The isolation tests are assumed to yield detection decisions in the absence of a failure with probability .047 and missed detection decisions in the presence of a failure with probability .02, each with hypergeometric holding time distributions that have means of approximately 3 samples.

Table 1 (derived from [22]) shows the results obtained for the state probability distribution obtained by an exact calculation of the transient solution and by use of the modified approximation just described. The approximation derived from the modified limit theorem in [25] leads to an approximate distribution with zero probability assigned to the transient states (1, 2, 4, 5, and 6). Results are shown for both values of the component failure probability, and, for one case, two different values for the number of samples K in the mission. The agreement is seen to be excellent in all cases. It should be noted that for the third case shown in Table 1, the modified limit theorem in [25] yields a probability of occupying state 8 after 10,000 samples of .39344 and a probability of occupying state 9 then of less than 10(-8). This means that the modified limit theorem results are more than 13% too high for state 8 after 10,000 samples and 5 orders of magnitude too low for state 9 at that time. The alternative method, on the other hand, produces less than 1% error in approximating both of these values. Note that 10,000 samples is half the MTBF for a component with probability of failure at each sample equal to 5 x 10(-5). Thus, the alternative method yields the most benefit when the mean time to interclass transitions is closer to the longest mean holding time within the classes, as we might expect.

3 Performance Evaluation Using Markov Models

Often, the designer or evaluator of a fault tolerant system design is interested not only in the reliability of the system but also its performance capability. This is particularly true of aerospace systems where performance is often the primary design goal and reliability is provided to ensure the satisfaction of the performance specifications. The performance of a system can be affected significantly by the occurrence of failure events and how the RM logic responds to these events. The Markov modeling method discussed above has the potential for evaluating fault tolerant system performance statistics where the effects of failures and RM decisions are included.

To clarify the meaning of performance evaluation in the context of fault tolerant system behavior, consider a simple example. Suppose we have two identical timepieces with known 1-sigma drift rates of σ and known failure rates, where a timepiece fails by stopping. Our objective is to measure the current time as accurately as possible. The strategy we use to achieve this (the RM strategy) is as follows. If we think both timepieces are working, we simply average their readings to generate our time estimate. To detect a failure (one timepiece stopped), we test every Δ hours to see if the difference between the indicated times exceeds a threshold that increases linearly in time. If such a detection is made, a third, outside time source (probably less accurate or more expensive) is used to determine which timepiece to declare as failed, and this third source is subject to random errors. A timepiece identified as failed is discarded (thus, this RM logic displays the irreversible decision characteristic discussed earlier).

Now, the performance question is this: After KΔ hours, how accurately do we know the time? Clearly, if both timepieces have failed by then, we do not know the time at all, and the probability of this situation is what traditional reliability analysis techniques would evaluate. However, if one timepiece is discarded on the basis of an RM decision and then the remaining one fails, then we also do not know the time at all. A Markov model of the system behavior that includes RM events would capture this situation as well as the previous one. If no timepiece failures occur and no false RM detections occur, then the averaging strategy produces a 1-sigma accuracy after $K\Delta$ hours of $\dfrac{K\Delta\sigma}{\sqrt{2}}$. Either a Markov model for the system behavior or a relatively simple reliability calculation shows that the probability of this situation is $(1-P_f)^{2K}(1-P_{fa})^K$.

Table 1 Comparison of exact and approximate distributions for SCDR system

	Exact results (state 1 to 9)	Approx. Results using step-size=50
k = 10,000	0.000027684938	0.000028037752
	0.000009194938	0.000008243412
$\varepsilon = 5 * 10^{-8}$	0.999432770000	0.999434580306
	0.000000000047	0.000000000021
	0.000000000003	0.000000000021
	0.000000000156	0.000000000241
	0.000061669460	0.000054278419
	0.000468653400	0.000474833582
	0.000000028905	0.000000026245
k = 15,000	0.000027582208	0.000027927991
	0.000009187481	0.000008233990
$\varepsilon = 5 * 10^{-8}$	0.999183050000	0.999184825377
	0.000000000057	0.000000000020
	0.000000000003	0.000000000020
	0.000000000156	0.000000000236
	0.000061654051	0.000053300697
	0.000718480390	0.000725668978
	0.000000044321	0.000000042689
k = 10,000	0.000033414812	0.000016518806
	0.000835702550	0.000016518806
$\varepsilon = 5 * 10^{-5}$	0.588035090000	0.588829863265
	0.000006652044	0.000000015147
	0.000000329155	0.000000015147
	0.000003095837	0.000000177613
	0.036292407000	0.039965861579
	0.353017460000	0.349653693105
	0.021775854000	0.021528998622

We have now answered the performance question for a few obvious scenarios. But what about all the other possibilities? Consider one particular case: One timepiece fails during the first time step and is not detected in the remaining time while the other timepiece works nominally the entire time. Because no times. Therefore, the average is biased by $\dfrac{(K-1)\Delta}{2}$ while the 1-σ variation

about that biased time estimate is $\dfrac{\sqrt{K^2+1}\,\Delta\sigma}{2}$, so the total root mean square

(rms) error in the time estimate is $\dfrac{1}{2}\sqrt{(K-1)^2\Delta^2 + (K^2+1)\Delta^2\sigma^2}$. This
possibility occurs with probability $P_f(1-P_f)^K(1-P_d)^{(K-1)}$.

Although it would be tremendously time-consuming for K any larger than a few samples, conceptually we could go through every possible failure and RM logic event history until we have exhausted them all. For each one, we would have a probability and a value of the performance measure, which in this case is rms accuracy of the time estimate. Thus, we would generate a pmf for the value of the performance measure, which can be used to calculate mean performance, performance variance, probability that performance exceeds (or does not exceed) a specified value, and so forth.

Our challenge then is to find a way of constructing the performance pmf without the impractical step of exhaustively evaluating every possible history of failure events and RM events. As we show in [23,24], a technique that we developed allows us to do this approximately with calculations that are typically no more cumbersome to implement than the solution for the transient history of a Markov model. This procedure is based on the "s-transform," which we explain below.

Suppose the behavior of the configuration of the system is described by a time-invariant Markov model with transition probability matrix Φ. Suppose also that the performance measure can be constructed as a sum of incremental values that are associated with each time step over the interval of operation, where the incremental values depend only upon the state that is occupied by the configuration process. In other words, assume that the performance measure J for a complete mission can be written as:

$$J = \sum_{k=1}^{K} J_k$$

where K is the total number of time steps in the mission and J_k is the incremental value of the performance measure contributed at time step k of the trajectory, where J_k can take N possible values corresponding to each of the possible states of the configuration. Let:

$$M(s,1) = [m_{ij}] = \phi_{ij}\exp(s \cdot J_i)]$$

where J_i is the incremental performance value corresponding to occupancy of state i for exactly one time step.

Consider now the product M(s,1) M(s,1). The (i,j) element of this product is given by:

$$\sum_{k=1}^{N} \phi_{ik} \exp(s \cdot J_k) \; \phi_{kj} \exp(s \cdot J_j) \; = \; \sum_{k=1}^{N} \phi_{ik} \; \phi_{kj} \exp[s(J_k + J_j)]$$

Each term in this sum is in the form of a probability times an exponential function with exponent equal to s times a performance value. The probability in this term is the probability of making the transition from state i to state j in two time steps through an intermediate transition to state k. The performance value in the exponent is the sum of the performance measure increments incurred on each of these steps (J_k on the first step to the intermediate state k plus J_j on the second step that transitions to state j). As a result, the matrix product M(s,1)M(s,1) completely characterizes the probabilities of every two-step transition and the cost associated with those two-step trajectories.

In similar fashion, let:

$$M(s,n) \; = \; [M(s,1)]^n$$

Then M(s,n) completely characterizes the trajectories of the system configuration over n time steps in terms of both probability and accumulated performance measure value. If π_0 is the initial probability distribution, then π_0 M(s,n) gives us the distribution of probabilities and cost values of all trajectories ending in each state at time step n. We denote this distribution as:

$$\pi_n^s(i) \; = \; \underline{ith} \text{ element of } M(s,n) \bullet \pi_0,$$

Then $\pi_n^s(i)$ characterizes the distribution of performance for trajectories ending in state i.

Now, it can easily be shown that $\pi_n^s(i)$ also generates the moments of the accumulated value of performance for each ending state over n time steps. Therefore, we also have:

$$\pi_n^s(i) \; = \; M_0^i + M_1^i \cdot s + M_2^i \cdot \frac{s^2}{2!} + \ldots$$

where M_k^i is the k-th moment of the performance value accumulated during the mission assuming that the configuration trajectory terminates in state i.

Truncating the power series representation of the exponential function $\exp(s \cdot J_i)$ at its kth term allows us then to directly compute approximate values for the moments of the accumulated performance measure value assuming termination of the process in each state.

The moments of the performance measure are important results. Obviously, one value of interest is the mean performance, and the calculation just discussed provides us with values for the mean performance given the terminating state. Solution of the original Markov model provides the probability of termination in each state, and these termination probabilities can be used to weight the individual performance mean values to determine the overall mean performance value. By considering alternative design philosophies or different values for the design parameters (such as the values of the thresholds used in the RM logic), the dependence of the mean performance for the system on these parameters or design forms can be found, thus providing an analysis tool for optimizing the design.

The discussion of the preceding paragraph can be repeated almost verbatim for any other moment of the performance value, including the mean square value.

Although the mean performance value and other moments of the performance value are useful results for judging the overall system performance, the designer of a fault tolerant aerospace system is more typically interested in the probability that the performance exceeds (or fails to exceed) a certain level. For evaluating this, knowledge of the performance value moments is insufficient. Instead, we need an approximate evaluation of the entire probability mass function of the performance value. From the moments that we have evaluated, we can get such an approximation.

Consider for the moment only one of the n sets of moments conditioned on the terminating state, say the set of moments M_n^i. The maximum entropy approximation to the pmf that produces these moments can easily be shown to be:

$$\text{pmf}(J|i) = \exp[a_{0i} + a_{1i} J + a_{2i} J^2 + \dots + a_{mi} J^m]$$

We then must find the values of the coefficients a_{ki} that correspond to the given moments M_n^i. This can be accomplished by a numerical optimization scheme, such as the steepest descent method.

To demonstrate the evaluation of performance pmfs directly from Markov behavior models, we applied the s-transform technique to a complicated aerospace mission, namely an inertial upper stage (IUS) test mission. The IUS is a space vehicle used to boost satellites from low earth orbits to higher orbits. A typical test mission consists of an 8 minute boost phase, followed by a long coast in the resulting orbit (typically about an hour), then an orbit transfer initiation burn followed by another long coast (usually several hours), then finally an orbit insertion burn at the new orbit.

We split the complete mission into these five phases and evaluate a separate s-transform model for each phase, using the conditions at the end of the previous phase as initial conditions. In this paper, we will present only results from the first burn phase. For results for the complete IUS mission, the reader is referred to [23].

The subsystem for which we are interested in the performance is the attitude rate measurement system. On the IUS, the attitude rate gyro system is a strapdown system with five gyros mounted such that their sensitive axes all lie along the surface of a cone with vertex half angle 54.735 degrees [26]. Because attitude rate is a three-dimensional vector quantity, the five gyros in a skewed configuration allows the unit to tolerate two failures before losing its ability to measure attitude rate.

The IUS attitude rate measurement system uses the parity vector approach to detect and isolate failures. Any detected failure is first isolated (instantaneously) to an adjacent pair of gyros, then further FDI tests isolate the failure to one of that pair. The RM logic is irreversible. The power supply to the gyros is also checked by a built-in test (BIT), which is also subject to error. Further details of the RM logic are given in [23,26].

Based upon the RM logic described above, the Markov model for the system configuration behavior has the nine states listed below:

Table 2. Markov Model States of the IUS Model

Model State	Description
0	Pentad, no sensor failures
1	Pentad, missed failure
2	Adjacent triad, false alarm
3	Adjacent triad, failure correctly isolated to pair
4	Adjacent triad, failure incorrectly isolated to pair
5	Quartet, failure correctly isolated to sensor
6	Adjacent triad, FDI terminated
7	Non-adjacent triad, power supply failure detected by BITE
8	Quartet or non-adjacent triad, one failed sensor in use
9	Vehicle loss

Each state definition includes the geometrical configuration of the gyros still in use and the fault detection system status. The transition probabilities for the

Markov model can be constructed from the values for the component failure rates, the RM test thresholds, and the covariance of the measurement noise. The details of this are given in [26].

The attitude rate measurement is obtained from the available sensors (five, four, or three) by a pseudoinverse calculation. Our performance metric is the mean square error (MSE) in the instantaneous attitude rate measurement, including any bias that might be present due to an undetected failure. We scale the units of the MSE in attitude rate such that the smallest performance value associated with any state of the Markov model has a value of unity and then round each value to the nearest integer so that the powers of s appearing in the s-transform evaluation technique are always integers. Upon doing this, we get the following values for the performance metric as a function of the state that is occupied for the dynamic phases (burns and attitude maneuvers) of the mission:

Table 3 Performance Values used in the IUS Model

State #	Performance Value
1	1
2	1
3	5
4	5
5	6
6	2
7	5
8	2
9	2
10	-

We then use the s-transform technique to calculate the pmf of the performance metric for termination in each state (except State 10, which is system loss). By weighting each of these performance value pmfs by the probability that we terminate the mission in this state and combining them, we get the complete performance value pmf for this phase of the mission.

We considered the IUS dynamic phase for a duration of 100 time steps. This does not necessarily correspond to the length of any of the dynamic phases of the IUS mission, but it provides an indication of the use of the s-transform method. For a mission of this duration, the Markov model yields a probability of system loss (i.e. probability of occupying State 10 at the end of the mission) of $2.563 \times 10^{(-6)}$.

Figures 4 and 5 show the performance value pmf that was constructed for this 100 time step IUS mission. Note that the vertical scale is a logarithmic scale. Figure 4 shows the pmf constructed from 5 coefficients determined from the moments generated by the s-transform method while Figure 5 shows the same pmf constructed from 7 coefficients. They are almost identical, but the 7-coefficient construction captures a bit more accurately the sudden drop in the pmf value for performance values around 210. Both pmfs indicate a large spike at the performance value of 100. This is due to the fact that the incremental performance value 1. is associated with States 1 and 2, and therefore occupancy of these two states for the entire mission leads to a cumulative value of the performance measure over the 100 time step mission of exactly 100, which is the minimum possible value. For the IUS Markov model used here, the probability of occupying these two states for the entire 100 time step mission is .999951, which is the value of the pmf spike at performance value 100.

With the pmf constructed as in Figures 4 or 5, we can answer questions regarding the probability of obtaining a particular value for the performance metric during this phase of the mission or the probability of exceeding a particular value. For instance, from either Figure, we see that the probability of exceeding a value of 210 for the performance metric is of order 10(-9), which we obtain by integrating the pmf beyond 210 on the horizontal scale.

Figure 4. IUS Dynamic Phase s-PMF, 100 TS, 5C, FLIMIT=1.E-8

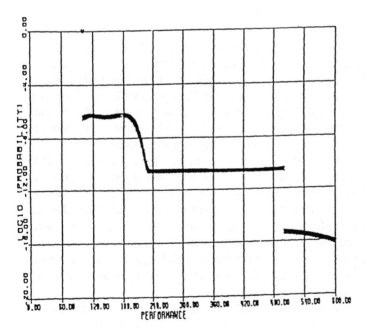

Figure 5. IUS Dynamic Phase S-PMF, 100 TS, 7C, FLIMIT=1.E-8

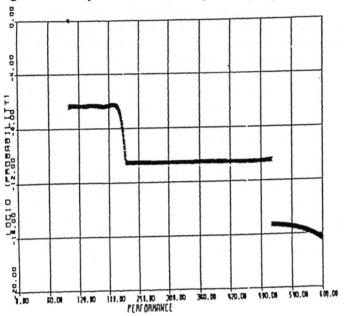

4 Conclusion

We have discussed various means by which Markov modeling has been applied to reliability and performance analysis for fault tolerant aerospace systems. The key quality of aerospace systems that make dynamic analysis methods such as the Markov modeling method necessary for their analysis is that they are highly automated and are usually required to react to changes in the system status very quickly. Because these responses can entail changes in the control and RM architectures, the configuration of the system itself becomes dynamic.

We have outlined how Markov modeling can be applied to evaluate the reliability and performance of these systems. Along the way, we have pointed out the difficulties attached to performing these types of analyses. The major difficulty stems from the large number of states that are typically needed to describe the system configuration, the long duration of operation for such systems relative to the time between RM decisions, the need to find transient solutions to the resulting models, and the fact that RM logic for these systems increasingly tends to involve sequential tests, which render the usual Markov assumption on the system behavior invalid.

We have briefly outlined how semi-Markov theory can be used to model system configuration behavior when the Markov assumption is violated. We have also discussed a limit theorem that allows large, computationally impractical models to be approximated by a much smaller aggregated model that behaves like a Markov process over long time periods. We have also discussed how to avoid one of the weaknesses of the limit theorem results by a useful approximation.

Finally, we have shown preliminary work from a method to evaluate the probability mass function of the performance measure of a fault tolerant system assuming that the performance measure can be divided into a sum of incremental terms that are a function only of the state being occupied.

As the missions for which aerospace systems are designed become more and more autonomous and as the computational capability of devices approved for real time, on-vehicle use expands, there is no doubt that fault tolerant aerospace systems will become widespread in the future. For the designer of such systems, a reliability or performance evaluation tool is invaluable in providing insight on whether a design perform well enough, and if not why it does not. In light of these developments, further work on evaluation tools that use dynamic strategies like Markov modeling will arise from necessity.

References

1. Walker, B.K., Wereley, N.M., Luppold, R.H., Gai, E. : Effects of redundancy management on reliability modeling. IEEE Trans. on Reliability, vol 38, no 4, 475-482 (1989)
2. Luppold, R.H.: Reliability and availability models for fault tolerant systems. M.S. thesis, Dept. of Aeronautics & Astronautics, Mass. Inst. of Techn. (1983)
3. Howard, R.A.: Dynamic probabilistic systems: Volume 1 - Markov models. New York: Wiley & Sons 1971
4. Parzen, E.: Stochastic processes. San Francisco: Holden-Day (1962)
5. Walker, B.K.: Reliability evaluation for fault tolerant systems by semi-Markov modeling. Submitted to Reliability Engineering and System Safety (1993)
6. Walker, B.K.: A semi-Markov approach to quantifying fault tolerant system performance. Sc.D. thesis, Dept. of Aeronautics & Astronautics, Mass. Inst. of Techn. (1980)
7. Beard, R.V.: Failure accommodation in linear systems through self-reorganization. Ph.D. thesis, Dept. of Aeronautics & Astronautics, Mass. Inst. of Techn. (1971)
8. Jones, H.L.: Failure detection in linear systems. Ph.D. thesis, Dept. of aeronautics & astronautics, Mass. Inst. of Techn. (1973)
9. Massoumnia, M.-A., Verghese, G.C., Willsky, A.S.: Failure detection and identification. IEEE Trans. on Auto. Control, vol 34, no 3, 316-321 (1989)
10. Chow, E.Y., Willsky, A.S.: Analytical redundancy and the design of robust failure detection systems. IEEE Trans. on Auto. Control, vol AC-29, no 4, 603-615 (1984)
11. Lou, X.-C., Willsky, A.S., Verghese, A.C.: Optimally robust redundancy relations. Automatica, vol 22, 333-344 (1986)
12. Frank, P.M.: Enhancement of robustness in observer-based fault detection. in R. Isermann (ed.), Fault Detection, Supervision and Safety for Technical Processes-SAFEPROCESS '91 (Baden-Baden), Intl. Fed. of Auto. Control, Pergamon Press, Oxford (UK), vol 1, 275-287 (1991)

13. Hou, M., Muller, P.C.: Design of robust observers for fault isolation. in R. Isermann (ed.), Fault Detection, Supervision and Safety for Technical Processes-SAFEPROCESS '91 (Baden-Baden), Intl. Fed. of Auto. Control, Pergamon Press, Oxford (UK), vol 1, 295-300 (1991)

14. Wald, A.: Sequential analysis, New York: Wiley, 1947; reprinted by New York: Dover 1973

15. Willsky, A.S., Jones, H.L.: A generalized likelihood ratio approach to the detection and estimation of jumps in linear systems. IEEE Trans. on Auto. Control, vol AC-21, no 1, 108-112 (1976)

16. Basseville, M.: Detecting changes in signals and systems - A survey. Automatica, vol 20, 387-404 (1988)

17. Chien, T.T., Adams, M.B.: A sequential failure detection technique and its application. IEEE Trans. on Auto. Control, vol AC-21, no 5, 750-757 (1976)

18. Speyer, J.L., White, J.E.: Shiryayev sequential probability ratio test for redundancy management. J. Guidance, Control, & Dynamics, vol 7, no 5, 588-595 (1984)

19. Howard, R.A.: Dynamic probabilistic systems: Volume 2 - Semi-Markov and Decision Processes. New York: Wiley & Sons 1971

20. Feller, W.: An introduction to probability theory and its applications (Volume 2), 2nd ed., New York: Wiley & Sons 1971.

21. Wereley, N.M., Walker, B.K.: Approximate evaluation of semi-Markov chain reliability models. Reliability Engineering and System Safety, vol. 28, 133-164 (1990)

22. Srichander, R., Walker, B.K.: An approximate algorithm for evaluating semi-Markov reliability models. Proc. of 1989 American Control Conf. (Pittsburgh), IEEE, New York (1989)

23. Missana, Walker, B.K.: Performance evaluation of fault tolerant systems with application to the IUS. Proc. of 1988 AIAA Guidance, Navigation, and Control Conf. (Minneapolis), AIAA paper 88-4110, AIAA, Washington (1988)

24. Wereley, N.M.: An approximate method for evaluating generalized Markov chain reliability models of fault tolerant systems. M.S. thesis, Dept. of Aeronautics & Astronautics, Mass. Inst. of Techn. (1987)

25. Walker, B.K., Srichander, R.: Approximate evaluation of reliability and related quantities via perturbation techniques. Final tech. report on AFOSR Grant 88-0131, Dept. of Aerospace Eng. & Eng. Mech., U. Cincinnati (1989)

26. Daly, K.C., Harrison, J.V., Gai, E.: Evaluation of the redundancy management for the IUS navigation system. vol. 1, R-1315, The Charles Stark Draper Laboratory, Cambridge, MA (1979)

Reliability Analysis Under Fluctuating Environment Using Markov Method

Balbir S. Dhillon

Engineering Management Program, Mechanical Engineering Department
University of Ottawa, Ottawa, Ontario K1N 6N5, Canada

Abstract. This paper presents three mathematical models developed by apply-ing the Markov technique. Model I represents a single unit system operating under two types of fluctuating weather. It means the operational system has one normal operating state and two mutually exclusive weather dependent states. Model II represents a motor vehicle operating in fluctuating weather (i.e., normal, abnormal). The vehicle can fail either due to a human error or a hardware failure in either of the two up states (normal weather, abnormal weather). Model III represents a human operator working under alternating stress environments. The operator may commit an error under normal or stressful conditions. Reliability analyses for all the three models are developed using the Markov method.

Keywords. Reliability, Markov method, environment, human error, human operator, probability

1 Introduction

In general reliability analysis of systems are performed under the assumption that their operating environments are steady. In real life, this assumption may not be true for some systems. For example, electrical systems operating in outdoor environment (weather subject to change - normal, stormy), motor vehicles operating under changing weather conditions, and pilots or motor vehi-cle drivers operating their systems under alternating weather or other fluctuating factors. The Markov technique is quite useful to conduct ana-lyses of systems operating under fluctuating environments. In order to demonstrate applicability of the Markov method, this paper presents three mathematical models to represent all the above cases. Some of the similar models may be found in [1]-[7]. Model I represents a single unit system operating under two types of fluctuating weather (e.g., snow storm/freezing rain, rain storm/thunder and lightening). It means that the operating system has one normal operating state and two mutually exclusive weather dependent states. In other words, the operating system alternates between these three states and it can also fail in either of the three states [6]. Model II [5] represents a motor vehicle operating in fluctuating weather (i.e., normal, abnormal). The operating vehicle can fail either due to a human error or a hardware failure in either of the two up states (normal weather, abnormal weather). The failed vehicle is repaired back into its operational states. Model III [4] represent a human operator working under alternating stress environments. Two examples of the tasks performed by the human operator under alternating stressful environments are piloting an aircraft and driving a motor vehicle. The person may commit an error under normal or stressful condition. Reliability analysis for all the three models are presented below using the Markov technique.

2 Model I

Notation

λ_j	system constant failure rate in state j $(j=0,1,2)$
α_1	constant weather alternating rate from system state 0 to state 1
α_2	constant weather alternating rate from system state 0 to state 2
γ_1	constant weather alternating rate from system state 1 to state 0
γ_2	constant weather alternating rate from system state 2 to state 0
$P_j(t)$	probability that the system is in state j at time t $(j=0,1,2,3)$
s	Laplace transform variable
$R(s)$	Laplace transform of system reliability
MTTF	Mean Time to Failure
$P_j(s)$	Laplace transform of the probability that the system is in state j at time t $(j=0,1,2,3)$

The system transition diagram is shown in Fig.1. The assumptions associated with the system are that the system failure and other occurrences are statistically independent and failure and weather alternating rates are constant.

Laplace transforms of the state probabilities associated with Fig.1 are as follows [6]:

$$P_0(s) = \left\{ \prod_{i=1}^{3}(s + a_i) - \alpha_1\gamma_1(s + a_3) - \alpha_2\gamma_2(s + a_2) \right\}^{-1} \left\{ \prod_{j=1}^{3}(s + a_j) \right\} \quad (1)$$

$$P_1(s) = \frac{\alpha_1}{s + a_2}P_0(s) \quad (2)$$

$$P_2(s) = \frac{\alpha_2}{s + a_3}P_0(s) \quad (3)$$

$$P_3(s) = \frac{\lambda_0(s + a_2)(s + a_3) + \lambda_1\alpha_1(s + a_3) + \lambda_2\alpha_2(s + a_2)}{s(s + a_2)(s + a_3)} \quad (4)$$

where

$$a_1 = \lambda_0 + \alpha_1 + \alpha_2$$
$$a_2 = \lambda_1 + \gamma_1$$
$$a_3 = \lambda_2 + \lambda_2.$$

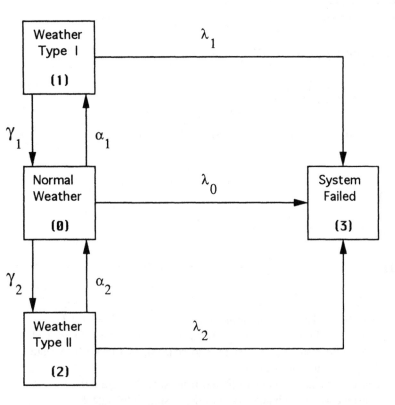

(0): outdoor equipment operates in normal environment

(1): outdoor equipment operates in type I abnormal weather state (e.g. snow storm, freezing rain)

(2): outdoor equipment operates in type II abnormal weather state (e.g., rain storm/thunder and lightening)

(3): system failed

Fig. 1 State-space diagram for Model I

The Laplace transform of system reliability is

$$R(s) = \sum_{i=0}^{2} P_i(s) = \frac{s^2 + sk_1 + k_2}{s^3 + s^2 w_2 + s w_1 + w_0} \tag{5}$$

where

$$k_1 = a_2 + a_3 + \alpha_1 + \alpha_2$$
$$k_2 = a_2 a_3 + \alpha_1 a_3 + \alpha_2 a_2$$
$$w_0 = a_1 a_2 a_3 - \alpha_1 \gamma_1 a_3 - \alpha_2 \gamma_2 a_2$$
$$w_1 = a_1 a_2 + a_1 a_3 + a_2 a_3 - \alpha_1 \gamma_1 - \alpha_2 \gamma_2$$
$$w_2 = a_1 + a_2 + a_3.$$

The system mean time to failure (MTTF) is given by

$$MTTF = \lim_{s \to 0} R(s) \tag{6}$$
$$= \frac{(\lambda_1 + \gamma_1)(\lambda_2 + \gamma_2) + \alpha_1(\lambda_2 + \gamma_2) + \alpha_2(\lambda_1 + \gamma_1)}{\alpha_1(u_1 + u_2) + \alpha_2(u_1 + u_3) + \lambda_0(u_1 + u_2 + u_3 + \lambda_1 \gamma_2)}$$

where

$$u_1 = \lambda_1 \lambda_2$$
$$u_2 = \lambda_1 \gamma_2$$
$$u_3 = \lambda_2 \gamma_1.$$

3 Model II

Notation

$P_i(t)$ probability that the vehicle is in state i at time t ($j=0,1,2,3$)

λ_i constant failure/human error rate associated with the vehicle: $i=1$ (state 1 to state 2), $i=2$ (state 0 to state 3), $i=3$ (state 0 to state 2), $i=4$ (state 1 to state 3)

α constant weather changeover rate from state 1 to state 0

β constant weather changeover rate from state 0 to state 1

μ_i constant repair rate of the vehicle: $i=1$ (state 3 to state 0), $i=2$ (state 2 to state 0), $i=3$ (state 2 to state 1), $i=4$ (state 3 to state 1)

s Laplace transform variable

$P_i(s)$ Laplace transform of the probability that the system is in state i at time t ($i=0,1,2,3$)

P_i steady-state probability that the system is in state i ($i=0,1,2,3$)

The system transition diagram is shown in Fig.2 [5]. The following assumptions are associated with the model:

(i)   Vehicle hardware and other failure rates are constant.

(ii)   Vehicle repair rates are constant.

(iii)   Failures are statistically independent.

(iv)   The weather alternating rates are constant.

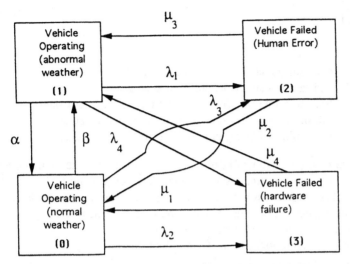

0: vehicle operates in normal weather   1: vehicle operates in stormy weather
2: vehicle failed due to a human error   3: vehicle failed due to a hardware failure

Fig. 2 Vehicle state-space diagram

The system steady state probabilities are as follows [5]:

$$P_0 = \frac{\mu_1 \mu_2 A_1}{A_1 \mu_1 \mu_2 + \beta \mu_1 \mu_2 + \lambda_1 \beta \mu_1 + \lambda_3 \mu_1 A_1 + \lambda_2 \mu_2 A_1 + \lambda_4 \beta \mu_2} \tag{7}$$

$$A_1 = \alpha + \lambda_1 + \lambda_4$$

$$P_1 = P_0 \frac{\beta}{A_1} \tag{8}$$

$$P_2 = P_0 \left\{ \frac{\lambda_1 \beta}{\mu_2 A_1} + \frac{\lambda_3}{\mu_2} \right\} \tag{9}$$

$$P_3 = P_0 \left\{ \frac{\lambda_2}{\mu_1} + \frac{\lambda_4 \beta}{A_1 \mu_1} \right\} \tag{10}$$

Setting $\mu_1 = \mu_2 = \mu_3 = \mu_4 = 0$ in Fig.2 we get the following expression for the system mean time to failure (MTTF):

$$MTTF = \lim_{s \to 0} R(s) = \lim_{s \to 0} \{P_0(s) + P_1(s)\} \tag{11}$$

$$= \left[\frac{\beta}{A_1} + 1 \right] A_1 \{(\beta + \lambda_2 + \lambda_3)A_1 - \alpha\beta\}^{-1}$$

4 Model III

Notation

λ_1 constant human error rate from state 0

λ_2 constant human error rate from state 2

α transition rate rate from normal state to the stress state

β transition rate rate from stress state to normal state

$P_i(t)$ probability of being in state i at time t $(j=0,1,2,3)$

The state-space diagram is shown in Fig.3 [3]. In this model, the human opera-tor performing the time-continuous task fluctuates between normal work and stress states. The task is associated with a system. The human error occurs from either the normal work state or the stress state (when the operator is performing under normal conditions or under stress).

The following assumptions are associated with this model:

1. Errors are statistically independent.

2. Human error rates are constant.

3. The human operator is performing a time-continuous task.

4. The rate of changing human operator condition from the normal state to the stress state and vice versa is constant.

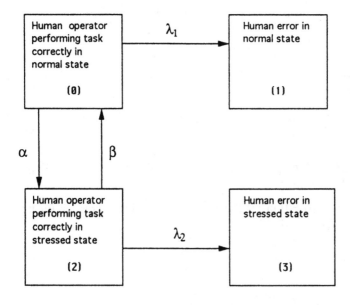

Fig. 3 State-space diagram (numerals denote corresponding states)

The time dependent state probabilities [8] associated with Fig. 3 are as follows:

$$P_0(t) = (x_2 - x_1)^{-1} [(x_2 + \lambda_2 + \beta)e^{x_2 t} - (x_1 + \lambda_2 + \beta)e^{x_1 t}] \tag{12}$$

where

$$x_1 = \frac{-b_1 + \sqrt{b_1^2 - 4b_2}}{2}$$

$$x_2 = \frac{-b_1 - \sqrt{b_1^2 - 4b_2}}{2}$$

$$b_1 = \lambda_1 + \lambda_2 + \alpha + \beta$$

$$b_2 = \lambda_1(\lambda_2 + \beta) + \alpha\lambda_2,$$

$$P_1(t) = b_4 + b_5 e^{x_2 t} - b_6 e^{x_1 t} \tag{13}$$

where

$$b_3 = \frac{1}{x_2 - x_1}$$

$$b_4 = \lambda_1(\lambda_2 + \beta)/x_1 x_2$$

$$b_5 = b_3(\lambda_1 + b_4 x_1)$$

$$b_6 = b_3(\lambda_1 + b_4 x_2)$$

$$P_2(t) = \alpha b_3(e^{x_2 t} - e^{x_1 t}) \tag{14}$$

$$P_3(t) = b_7[(1 + b_3)(x_1 e^{x_2 t} - x_2 e^{x_1 t})] \tag{15}$$

where

$$b_7 = \lambda_2 \alpha/x_1 x_2.$$

The human operator reliability is given by

$$R(t) = P_0(t) + P_2(t). \tag{16}$$

The mean time to human error is given by

$$MTTHE = \int_0^\infty R(t)dt = \int_0^\infty [P_0(t) + P_2(t)]dt = (\lambda_2 + \alpha + \beta)/b_2 \qquad (17)$$

References

1. Endrenyi, J.: Reliability modelling in electric power system. New York: John Wiley & Sons 1978

2. Dhillon, B.S.: Power system reliability, safety and management. Ann Arbor, Michigan: Ann Arbor Science/The Butterworth Group 1983

3. Dhillon, B.S.: Human Reliability With Human Factors. New York: Pergamon Press 1986

4. Dhillon, B.S.: Stochastic Models for Predicting Human Reliability. Microelectronics and Reliability, 22, 491-496 (1982)

5. Dhillon, B.S.: RAM Analysis of Vehicles in Changing Weather. Proceedings of the Annual Reliability and Maintainability Symposium, 48-53 (1984)

6. Dhillon, B.S., Natesan, J.: Stochastic Analysis of Outdoor Power Systems in Fluctuating Environment, Microelectronics and Reliability, 23, 867-881 (1983)

7. Dhillon, B.S., Rayapati, S.N.: Reliability Evaluation of Outdoor Electric Power Systems, International Journal of Energy Systems, 7, 85-89 (1987)

8. Dhillon, B.S., Rayapati, S.N.: Stochastic Behaviour of Man-Machine Systems Operating Under Different Weather Conditions, Microelectronics and Reliability, 26, 123-129 (1986)

Automatic Generation of Dynamic Event Trees: A Tool for Integrated Safety Assessment (ISA)

José M. Izquierdo, Javier Hortal, Miguel Sánchez, Enrique Meléndez

Consejo de Seguridad Nuclear, 28040 Madrid, Spain.

Abstract. The concept of Integrated Safety Assessment (ISA) is described in mathematical terms as the basis of a tool for its practical application to high risk installations with aggressive protections. It incorporates the dynamics of the facility as well as the operating environment, both subject to transitions between different time evolutions due to failures and/or system/operator interventions, with emphasis in deterministic transitions. The methodology can be considered an extension of PSA and accident analysis techniques that replaces the static event tree with a generalized dynamic event tree concept based on the theory of probabilistic dynamics. It is particularly suited to assess software (logic) aspects of protection systems, and can be of particular interest to regulatory agencies.

Keywords. Integrated Safety Analysis, Event Trees, Design Basis Accidents, Dynamic Reliability, Technical Specifications, Probabilistic Safety Analysis, Accident Management, Emergency Procedures, Protection Systems.

1 Introduction

This paper makes an attempt to formulate in a mathematical (and thus, precise) language the concept of integrated safety assessment (ISA) which can be applied for the design and assessment of software (logic) aspects of protection systems, focusing more in the assessment side. By software of the protection systems we understand the set of protection initiation signals with associated alarms, precautions, limitations of the protected system operation as well as fault, emergency and accident management procedures.

Software aspects of automatic protection systems have been designed within the nuclear community ((AEC 1972), (Almaraz 1986)), by defining a set of "design basis events", simulating their transient history under certain assumptions and trying to demonstrate that for these scenarios the facility will not violate prespecified damage limits. Failure probabilities are indirectly considered, often only in a qualitative sense, by imposing different damage limits to design basis events with probability of occurrence lying in different ranges (ANSI 1983).

Although the incorporation of reliability techniques allows a more detailed evaluation of these failure rates, the rest of the analysis is deterministic in nature, and reflects an extension of control theory methods, where it is also usual to define design basis events like step and ramp perturbations, in spite of the essential differences in role between both types of systems. A similar approach was taken for the EOP (Emergency Operating Procedures) design of the manual software (Speis 1985).

Soon after the occurrence of the TMI accident, this approach has been supplemented with many PSA studies where a generalized use of the fault tree-static event tree technique, attempts to identify significant scenarios involving multiple failures outside the design scope. However, the static nature of the methodology (US NRC 1983) emphasizes only the hardware aspects and has been criticized (see the overview of (Siu 1992)), particularly when operator interventions are taken into account.

Indeed, the influence of the software of the protection systems in static event trees is obscured in the current methods because, among others (Kafka 1992), implicit assumptions are made about the following aspects:

a) When system unavailability or transition probabilities are given in "per demand" terms, the probability of the demand itself, i.e. of the triggering of system/operator intervention conditions, should also be ascertained. In fact, with the exception of stochastic transitions, like instrument failures/operator errors, the intervention will only occur if automatic or operator initiation criteria are satisfied.

b) Some component failure rates, or failure upon demand, are expected to be dependent on process variables, for instance as a result of instrument out of range failures. This particularly applies to human interventions.

c) In most descriptions, a given damage, as for instance reactor core melting, is actually considered a necessary consequence of a given combination of failures. Although a verification of the assumptions made to correlate damage and probability should be performed by simulating the dynamic response of significant sequences, it is difficult to ensure the consistency of the overall approach (see comments at the end of Section 3.3 below).

In order to incorporate the assessment of these software aspects in a suitable probabilistic approach, an effort was initiated in 1987 to develop a more consistent and comprehensive methodology ((Izquierdo et al. 1988) and (Izquierdo, Sánchez-Perea 1990)) based on a general licensing philosophy (Izquierdo, Villadóniga 1978). It was concluded that, in developing event trees, coupled dynamic simulation of the plant response was very convenient to assess whether any mitigating system was actually demanded, incorporating thus the software aspects of automatic systems as well as the operator/crew interventions into the risk evaluations.

While the practicality of such a concept for large facilities was in question, the maturity of the situation today makes it feasible. In fact, this coupling between simulation and probabilistic techniques can be fulfilled by the so called "Dynamic Event Trees" (DET). The approach can be extended to any complex installation where protection optimization is required, and its feasibility has been already tested (Izquierdo, Sánchez Perea 1990) with the TRETA (Meléndez 1992)/DYLAM (Cojazzi et al. 1992a) package, as a result of a joint cooperation between the TRETA simulator development team of CSN (Izquierdo et al. 1987) and the DYLAM DET generator team of JRC-ISPRA (Amendola, Reina 1981). Other organizations are also using dynamic methods for the generation of DETs, for instance building up and applying the theory of probabilistic dynamics and continuous event trees (Devooght, Smidts 1992ab), discretizing the process variables and using Markov techniques (Aldemir 1991) or comparing DYLAM-like DET results with conventional PSA studies (Acosta, Siu 1991).

The major drawback of these attempts, which is today the subject of considerable research, is how to prevent an excessive number of sequences to analyze (see for instance (Marseguerra, Zio 1992) and (Cojazzi et al. 1992b). Realistic simulations of large facilities require large number of process variables and become a potential limiting factor. Moreover, operator models involving a wide range of possible errors strongly increase the stochastic nature of the transitions, making the problem easily unmanageable (Siu 1992).

Considerations about simultaneous failures, modes of operation of the components, emphasis on failures/errors on demand, etc. allow the reduction of the number of branches (Cacciabue et al. 1992). These, together with preliminary ideas about ISA have also been addressed (Cojazzi et al. 1992b).

We support the idea (Macwan et al. 1991), that an intelligent approach to the model of the operator strongly reduces the size of the stochastic problem, emphasizing the importance of transitions upon deterministic demand which, as shown below, do not increase the number of sequences over the static PSA. Additionally, the objective of the study may be limited to check some of the overall DBEPs (see Section 4 below) with also drastic cuts in the number of sequences.

The rest of the paper is organized as follows: In Section 2 we introduce the precise definition of the terms used in the next sections. Section 3 deals with the mathematical description of the dynamic and deterministic event trees, while in Section 4 the ISA methodology is described, focusing in the kind of problems accounted for, the steps of the methodology and the tools available for its implementation.

2 Definitions

In this Section we define the terms used later, to ensure precise meaning. Without loss of generality, and to simplify the notation, we do not consider operator states. In fact, operator state extensions of the theory of probabilistic dynamics are available (Devooght, Smidts 1992c), and (Macwan et al. 1991) already show a useful practical approach.

2.1 Description of the Facility

Facility: An installation and its protections, i.e. $I + P_I$.

Installation, I: Set of interrelated systems to obtain a benefit by performing several operating functions. For instance, the control system is an Installation system.

System, S: Set of: i) components, ii) articulated in a certain way, iii) with characteristics and iv) specifications to perform a certain system function as a part-task of the facility functions.

Protection associated to I, P_I: Set of protections, i.e. systems and/or system features (i), ii), iii), iv) in the system definition) that prevent the facility from causing any unacceptable damage by performing several protective functions. For instance, containment wall painting is a protection of Nuclear Power Plants.

Conservative protection: A protection that includes a sufficient one.

2.2 Description of the Protection System

Protection System: Set of devices/interventions that generate the signals for automatic/manual actuation of protections. These will include some or all of the following:

Limitation system: Function, to generate signals triggering protective actions without stopping I, i.e. drive it into a stable state in some operating mode.

Trip system: Function, to generate signals to quickly stop I, i.e. drive it into a stable state in some non operating mode.

Safeguard initiation system: Function, to generate signals triggering protective actions once I has been stopped. In conjunction with trip and limitation systems, drive it into a safe state (see Section 4.1.c) in some non operating mode.

Manual Protection System: Set of rules and operator/crew support systems/organizations that determine/help safety oriented operator/crew actions. For instance, several abnormal operating procedures are a part of the manual limitation system of a NPP.

Automatic Protection System: Set of devices that automatically trigger protections.

2.3 Description of the Facility States

Combined state (or state) of a system of N components:

Set of combined states, $(j, \vec{x})$, where j, the system status vector, may change according to a probability of transition and $\vec{x}$ denotes the process vector.

Steady[1] state: Time independent state $(\alpha, \vec{x}_0)$. Note that there may be incompatibility of certain plant status with steady process states.

Stable state: A steady process state such that, if subject to a small perturbation, the system or installation comes back to that state once the perturbation dies out.

Safe state: See Section 4.1.c.

Plant/system status vector: Integer vectors j describing the different states of facility components, (nominal, derated, failed in a given mode, etc.),

$$j = (j_1, j_2,..., j_N)$$

Process vector: Real vectors $\vec{x}$ describing the facility process evolutions (temperatures, flows, etc.),

$$\vec{x}(t) = (\vec{x}_1(t), \vec{x}_2(t),..., \vec{x}_N(t))$$

Safety variable: A function $h_f(\vec{x})$ of the process state used to formulate initiation criteria of protections. See Section 3.3.

Damage variable: A function $D_i(\vec{x}, t)$ of the process state, used to formulate DBEP$_i$ damage limits for variable i. See Section 4.1.a.

Design Basis Envelope, DBEP: See Section 4.1.b.

2.4 Description of events, faults and transitions

System transient or system dynamics: Non steady time evolution of system process states, governed by a dynamic equation associated to a system status (see eq. (1)).

Driver transient: First transition.

Event: Instantaneous change in plant status vector.

Active (Passive) Event: Event triggering (not triggering) a transient.

Initiating Event: First active event.

Single event: Event involving changes in a single component.

Sequence of events: Time history of the plant status vector.

Path of a sequence: Time history of the combined state.

Top sequence: See Section 4.1.c.

Design Basis Event: See Section 4.2.

Fault: Event implying degradation of system functions.

Common Failure (or Common Event): Non single simultaneous fault (or event).

Transition: A change in the system dynamics.

Perturbation: A transition from a steady state.

---

[1]Greek/latin letters denote steady/transient states.

3 General Equation of Dynamic Event Trees. Deterministic Event Trees (Devooght, Izquierdo)

The relationship between deterministic (i.e. transitions upon deterministic demand) dynamic and classical event trees can be realized better if we write the general equations of probabilistic dynamics (Devooght, Smidts 1992c) in an alternate form, that we describe in Section 3.1 below. In Section 3.3 we show that the static event trees are a limiting case of deterministic event trees. This demonstrates the consistency of the theory of probabilistic dynamics, and shows that deterministic dynamic event trees are a valid generalization of the classical techniques.

3.1 Alternate Form of Probabilistic Dynamics: The General Equations of Dynamic Event Trees

The following form of the general integral equations of probabilistic dynamics is equivalent to the standard ones.

Let $\vec{g}_j(\tau, \vec{u})$ denote the equation of movement of a system along a dynamics j with initial conditions $\vec{u}$, i.e.

$$\vec{x} = \vec{g}_j(\tau, \vec{u}) \tag{1}$$

Let $\tau_j(\vec{u} \rightarrow \vec{x})$ be the time necessary to go from $\vec{u}$ to $\vec{x}$ flying along the path determined by the dynamics j. From eq. (1),

$$\tau = \tau_j(\vec{u} \rightarrow \vec{x}). \tag{2}$$

Let $p_{k \rightarrow j}(\vec{x})d\vec{x}$ be the probability of having a transition from dynamics k to dynamics j between $\vec{x}$ and $\vec{x} + d\vec{x}$.

We want to find the equations governing $\pi_j^n(\vec{x}, t)$, the probability density[2] of state $(j, \vec{x})$ at time t, after n transitions. Written without superindex n will mean the probability with any number of transitions, i.e. the sum for all n.

Define

$$\xi_j(\vec{x}, t) \equiv \sum_k p_{k \rightarrow j}(\vec{x}) \pi_k(\vec{x}, t) \tag{3}$$

where $\xi_j(\vec{x}, t)d\vec{x}$ is the probability density that the dynamics j starts at $(\vec{x}, t)$, as

---

[2] i.e. $\displaystyle\sum_j \int d\vec{x}\ \pi_j(\vec{x},t) = 1$

a result of previous transitions. Written with superindex n will mean the same but after n previous transitions.

Also define

$$P_{j-k}(\vec{u},\ t) \equiv e^{-\int_0^t p_{j-k}(\vec{\xi}(s,\vec{u}))\left|\frac{d\vec{x}}{ds}\right| ds} \tag{4}$$

where P_{j-k} is the survival probability to transitions $j{\to}k$ while flying along dynamics j during a time t. Clearly,

$$P_{j-k}(\vec{u}{\to}\vec{x}) \equiv P_{j-k}\left(\vec{u},\ \tau_j(\vec{u}{\to}\vec{x})\right) \tag{5}$$

is the survival probability to transitions $j{\to}k$ while flying along dynamics j from $\vec{u}$ to $\vec{x}$. Note that if $\vec{x}$ cannot be reached from $\vec{u}$, $\tau(\vec{u}{\to}\vec{x})$ is ∞ and $P_{j-k}=0$.

Then,

$$P_{j*k}(\vec{u}{\to}\vec{x}) \equiv \underset{\substack{l \\ j*l \\ l*k}}{\Pi} P_{j-l}(\vec{u}{\to}\vec{x}) \tag{6}$$

is the survival probability to transitions other than $j{\to}k$ while flying along dynamics j, from $\vec{u}$ to $\vec{x}$, and

$$P_j(\vec{u}{\to}\vec{x}) \equiv P_{j-k}(\vec{u}{\to}\vec{x})\ P_{j*k}(\vec{u}{\to}\vec{x}) = \underset{\substack{l \\ j*l}}{\Pi} P_{j-l}(\vec{u}{\to}\vec{x}) \tag{7}$$

is the survival probability to any transition while flying along dynamics j from $\vec{u}$ to $\vec{x}$. We finally define,

$$f_{j-k}(\vec{u}{\to}\vec{x})\left|\frac{d\vec{x}}{dt}\right| \equiv p_{j-k}(\vec{x})P_{j-k}(\vec{u}{\to}\vec{x})\left|\frac{d\vec{x}}{dt}\right| = -\frac{\partial}{\partial t}[P_{j-k}(\vec{u},\ t)]_{t=\tau(\vec{u}{\to}\vec{x})} \tag{8}$$

where $f_{j-k}(\vec{u}{\to}\vec{x})d\vec{x}$ is the probability of starting dynamics k at $(\vec{x},\ t)$ after surviving to transitions $j{\to}k$ from $\vec{u}$ to $\vec{x}$.

With all these definitions, it is clear that the following iterative equation links ξ_j^n between themselves:

$$\xi_j^n(\vec{x},\ t) = \theta(t)\underset{\substack{k \\ k*j}}{\sum} \int d\vec{u}\ f_{k-j}(\vec{u}{\to}\vec{x})P_{k*j}(\vec{u}{\to}\vec{x})\xi_k^{n-1}(\vec{u},\ t-\tau_k(\vec{u}{\to}\vec{x})) \tag{9}$$

The relation between ξ and the solution π can be obtained through

$$\pi_j^n(\vec{x},\ t)=\int d\vec{u}\ P_j(\vec{u}\rightarrow\vec{x})\xi_j^n(\vec{u},\ t-\tau_j(\vec{u}\rightarrow\vec{x})) \tag{10}$$

and the system is closed usually by the initial and boundary conditions (Section 3.2)

$$\xi_j^0(\vec{x},\ t)d\vec{x}=\delta_{j_0}\delta(\vec{x}-\vec{x}_0)\pi_{\alpha j_0}^0(\vec{x}_0)\delta(t)dt$$

$$\pi_{\alpha j_0}^0(\vec{x}_0)\equiv\pi_\alpha^0\ X_{\alpha j_0}(\vec{x}_0) \tag{11}$$

where $(\alpha,\ \vec{x}_0)$ is an initial steady state with probability π_α^0 (i.e. $\vec{g}_\alpha(t,\ \vec{x}_0)$ is time independent; see Section 3.2 below), and the initiating event, described by $\alpha\rightarrow j_0$ in a deterministic way[3], introduces the dynamics j_0 in the system (no longer in steady state, driver transient) with a probability $X_{\alpha j_0}(\vec{x}_0)$ at time zero (see eq. (12)).

Equations (9), (10) and (11) can be considered as a mathematical description of a dynamic event tree. Note that this approach, although rigorous, requires the evaluation of integrals in many dimensions and through many dynamics, a task that is extremely difficult to accomplish unless the domain of integration is restricted. One practical procedure is the so called event method, which establishes some rules (branching and sequence expansion rules in (Siu 1991)) to decide a discrete number of paths for which the contribution is evaluated. As will be seen below, the case of transitions upon deterministic demand inherently reduces to such a case and there is no need to use branching and/or expansion rules.

3.2 The Case of Steady Process Dynamics

We assume that the installation is in state $(\beta,\ \vec{x}_0)$ with $\vec{x}_0$ a steady state, immediately after a maintenance operation that takes place at a time -T. We select time 0 at the time of occurrence of the first transition. A number of passive events may occur between -T and 0, that only involve the steady states. During this period, the transition probabilities become time independent, and $P_{\beta-\alpha}(t)$ becomes the classical survival probability. It is easy to show (Devooght, Smidts 1992c) that eqs. (9), (10), if used to obtain the probability π_α^0 of the state $(\alpha,\ \vec{x}_0)$ at time zero become equivalent to the Markovian equations of standard reliability (Bharucha-Reid 1960). Once found, this steady state, initial condition of the active phase, defines a facility state with several components eventually out of service.

---

[3] $\alpha\rightarrow j_o$ is assumed not to occur again during the time of interest of the study. See (Devoogth, Izquierdo) for a better description.

Once the active transition starts, it is assumed that the probability of further installation failures is small during the time the protective devices take to cope with the transient until a new steady state is reached. Thus after time 0, transitions j→k involve mostly the protection components and are dominated by deterministic demand.

3.3 The Case of Transitions Upon Deterministic Demand. Mostly Deterministic Dynamic Event Trees

We will assume that a subset $\{p_{i'-j'}\}$ are transitions upon deterministic demand. The demand is triggered if anyone of the setpoints described by the conditions of the safety variables $h_l(\vec{x}) = 0$, $l = 1, ..., N_{sp}$ are reached.

It is not difficult to show (Devooght, Izquierdo), taking into account eq. (8), that

$$f_{j'-k}(\vec{u} \to \vec{x}) d\vec{x} = \delta(\vec{x} - \vec{x}_s) X_{j'k}(\vec{x}_s) \theta(\tau_j(\vec{u} \to \vec{x}) - \tau_j(\vec{u} \to \vec{x}_s)) d\vec{x}_s$$
$$= \delta(t - t_s) X_{j'k}(\vec{x}_s) \theta(t - t_s) dt \tag{12}$$

where $X_{j'k}(\vec{x}_s)$ is the probability of the demand at the dynamic conditions $\vec{x}_s$ at the time t_s the demand arises (see below). Eq. (9) can then be simplified and the deterministic transitions be integrated giving

$$\xi_j^n(\vec{x}, t) = \delta(\vec{x} - \vec{x}_s) \sum_{k'} X_{k'j}(\vec{x}_s) \int d\vec{u} \; P_{k',j}(\vec{u} \to \vec{x}_s) \xi_{k'}^{n-1}(\vec{u}, \; t - \tau_k(\vec{u} \to \vec{x}_s)) \tag{13}$$

+ *stochastic contributions*

As a result, we see that the effect on the probability density (eq. (10)), of the contribution to $\xi_j^n(\vec{x}, t)$ (defined in eq. (9)) of the deterministic transitions, is given by

$$\pi_j^n(\vec{x}, t) = \sum_{k'} \int_{\Gamma_j(\vec{x}, t)} d\vec{u}_s \; X_{k'j}(\vec{u}_s) P_j(\vec{u}_s \to \vec{x}) *$$

$$\int d\vec{u} \; \delta(\vec{x} - \vec{g}_j(t - \tau_k(\vec{u} \to \vec{u}_s), \; \vec{u}_s)) \; P_{k',j}(\vec{u} \to \vec{u}_s) \xi_{k'}^{n-1}(\vec{u}, \; t - \tau_k(\vec{u} \to \vec{u}_s) - \tau_j(\vec{u}_s \to \vec{x})) \tag{14}$$

+ *stochastic contributions*

i.e. the domain of integration of eq. (10) is restricted to surfaces $\Gamma_j(\vec{x}, t)$, induced by the initiation conditions and the dynamics j through

$$x_s \epsilon \Gamma_j(\vec{x}, t) \qquad \Gamma_j \equiv \overset{N_{sp}}{\underset{l=1}{U}} \Gamma_{jl}(\vec{x}, t)$$

(15)

$$\Gamma_{jl} \left\{ \begin{array}{l} h_l(\vec{x}_s) = 0 \\ \tau_j(\vec{x}_s \rightarrow \vec{x}) < t \end{array} \quad l=1,..., N_{sp} \right\}$$

Note that if there are several initiation conditions, i.e. several l values (like for instance different trip functions), the union of all individual surfaces would be the one to consider. Equations (13), (14), (15) describe the mostly deterministic dynamic event trees.

3.4 Relationship Between Deterministic DETs and Classical (Static) Event Trees

Finally, in the case where all the transitions are upon demand, and independent of each other, all $P_{j \rightarrow k}$ and $P_{j \neq k}$ of eqs. (13), (14), are 1 between the setpoint surfaces, and the initial conditions and initiating event usually will determine, at the first transition (if any), a single point of the $\Gamma_j(\vec{x}, t)$ surfaces, which is the intersection of the initial dynamic path, j_0, followed by the system and the first setpoint surface crossed.

From there, any new dynamic path will also determine usually unique points, $\vec{u}_s^n$, and then all the integrals of eqs. (13), (14) reduce to the values of the integrands at the corresponding surface setpoint intersections, $\vec{u}_s^n$. That is, combining eq. (10) with eq. (11),

$$\pi_j^0(\vec{x}, t) = \delta_{jj_0} \, \delta\left(\vec{x} - g_{j_0}(t, \vec{x}_0)\right) \pi_{\alpha j_0}^0(\vec{x}_0).$$

(16)

From eq. (13),

$$\xi_j^1(\vec{x}, t) = \pi_{\alpha j_0}^0(\vec{x}_0) \, X_{j_0 j}(\vec{u}_s^1) \, \delta(\vec{x} - \vec{u}_s^1) \, \delta\left(t - \tau_j(\vec{x}_0 \rightarrow \vec{u}_s^1)\right)$$

(17)

and from eq. (14),

$$\pi_j^1(\vec{x}, t) = \pi_{\alpha j_0}^0(\vec{x}_0) \, \delta\left(\vec{x} - \vec{g}_j(t - \tau_{j_0}(\vec{x}_0 \rightarrow \vec{u}_s^1), \vec{x}_0)\right) X_{j_0 j}(\vec{u}_s^1).$$

(18)

Iterating, we obtain:

Define

$$Path_n \equiv \left\{ \begin{array}{c} j_0 \rightarrow j_1 \rightarrow \ldots \rightarrow j_n \rightarrow j \\ \vec{x}_0 \rightarrow \vec{u}_s^1 \rightarrow \ldots \rightarrow \vec{u}_s^{n+1} \rightarrow \vec{x} \end{array} \right\} \tag{19}$$

$$\tau_n \equiv \tau_n(\vec{x}_0) \equiv \tau_{j_0}(\vec{x}_0 \rightarrow \vec{u}_s^1) + \sum_{l=1}^{n} \tau_{j_l}(\vec{u}_s^l \rightarrow \vec{u}_s^{l+1})$$

then

$$\pi_j^n(\vec{x}, t) = \pi_{\alpha j_0}^0(\vec{x}_0) \sum_{Path_n} \delta\left(\vec{x} - \vec{g}_j(t - \tau_n, \vec{u}_s^n)\right) \prod_{l=0}^{n} X_{j_l j_{l+1}}(\vec{u}_s^{l+1}). \tag{20}$$

Finally the probability at time t irrespective of $\vec{x}$ and n is given by

$$\pi_j(t) \equiv \int_0^{\infty} \pi_j(\vec{x}, t) d\vec{x} =$$

$$= \pi_{\alpha j_0}^0(\vec{x}_0) \sum_{n=0}^{\infty} \sum_{Path_n} \theta(t - \tau_n) \prod_{l=0}^{n} X_{j_l j_{l+1}}(\vec{u}_s^{l+1}) \tag{21}$$

with $\quad \vec{u}_s^{n+2} \equiv \vec{x}; \quad j_{n+1} \equiv j.$

If the initial conditions and initiating events/transitions are such that the system crosses all surfaces of the setpoints, eq. (21) when $t \rightarrow \infty$ is the sum of products of the probabilities upon demand found through the different paths. This result is the standard method used in static event trees, which actually calculates $\pi_j^n(t)$ where n is the expected number of transitions that may produce a given damage. However, if $\tau_j(\vec{u}_s^l \rightarrow \vec{u}_s^{l+1})$ is infinity for any j in any setpoint surface of a path, the contribution of this path to π^n would be zero indicating a wrong initial assumption concerning the expected transitions for this initial conditions and initiating events.

This conclusion clearly illustrates the influence of the design of the setpoint surfaces and initial conditions $(\alpha, \vec{x}_0)$ (i.e. the software of the protection system) in the event trees: They shape the $\theta(t - \tau_n)$ factors, and determine the dynamic conditions, $\vec{u}_s^l$ at the demand time t_s^l. Moreover, as shown in Section 4 the risk requirements force to correlate local damage and probability density, and it is very difficult to exclude paths in an a priori form. If a dynamic analysis of each sequence is performed in a decoupled way, the effort would be equivalent to a deterministic dynamic event tree evaluation, except that the triggering of the setpoints ought to be confirmed, something not intuitive at all and that risks to require many trials. In fact, such an extensive dynamic work is seldom performed in any PSA study and would be impractical in an ISA. Because the number of sequences is the same and the dynamic event tree computes all damage conditions at the same time as the probabilities, it seems the appropriate method to follow, particularly if the tools used are also able to handle discrete but stochastic transitions at least to the extent necessary to incorporate operator behavior.

4 The ISA Methodology

4.1 Risk Requirements for Software Design: The DBEP Concept

The following requirements are consistent with US nuclear industry practice and licensing regulations, but are formulated in a way applicable to any facility.

a) Selection of Damage Variables: Process variables characterizing the undesired situations shall be selected; in the nuclear industry, they are usually those indicative of barrier failures (Izquierdo, Villadóniga 1978) (e.g. for the design of automatic actions: fuel temperature, onset of boiling crisis, fuel cladding temperature, primary and containment pressure; for the design of emergency operating procedures: the critical safety functions).

b) Design Basis Envelope Plane (DBEP) requirements: Limits separating the acceptable and unacceptable conditions, described as regions in the plane probability-damage variables, shall be established and proved to satisfy and include all relevant regulatory constraints (Izquierdo, Villadóniga 1978) i.e.

$$DBEP_i \equiv \pi^L(D_i) \quad with \quad i=1, 2,..., n_{DBEP}$$

The envelope DBEP is then a set of regions in n_{DBEP} planes.

c) Top sequences and safe states requirements: A sequence that contains at least one point in the unacceptable region of at least one design basis envelope plane is a top sequence. A safe state is a stable state from which any initiating event will not produce any top sequence i.e. unacceptable damage. Every operating steady state shall be a safe state.

4.2 The Design Problem

Given an installation, I, find an optimum software of the protection P_I, such that:
  a) satisfies the risk requirements of Section 4.1.
  b) the set of safe states is maximum.
  c) the distribution among each type of manual and automatic actions (either limitation, trip and safeguard) optimizes protection interventions with a minimum of the more aggressive.
  d) the cost-benefit for facility production is optimum.

The complexity of this problem in large facilities forces sophisticated design methodologies (for example, the Revised Thermal Design Procedures of Westinghouse), strongly dependent on the details of the technology. For example, artificial events (Design Basis Events) with many bounding distortions of reality are used to obtain expected conservative protections.

4.3 The Assessment Problem

Given a facility, vérify that the software of the automatic and manual protection systems satisfies aspects a) and c) of the design problem, independently of the design method followed (for instance, with no reference to the design basis events). This is the precise purpose of the ISA method.

4.4 Steps of the Method

Being substantially an extension of the PSA method (US NRC 1983), the steps are similar. If applied in the same nuclear environment, ISA will benefit from PSA results on the one hand, and available accident analysis quick tools and associated models of the facilities (including operators) on the other hand. The steps simply follow in detail the logic flow required to feed eqs. (13), (14). A single/double asterisk will indicate that the problem involved is the same as in conventional PSAs/Accident Analysis and therefore appropriate assumptions and data are available.

Step 0 Selection of the objective: Define the DBEPs to be considered. This selection strongly conditions the remaining steps.

Step 1 Selection of initial states: Define the states at time 0 (α, $\vec{x}_0$).

  Step 1.1: (*) Select α which accounts for the previous passive events $\beta \rightarrow \alpha$ from -T. It should cover systems out of service either deliberately or failed.

  Step 1.2: (**) Select $\vec{x}_0$. Possible steady states are dependent on α, and are determined by a set of degrees of freedom (Izquierdo, Van Hoënacker 1992) sensible to the status of the control systems (loops in manual, etc.).

  Step 1.3: (*) Evaluate π_α^0. Note that it may depend on $\vec{x}_0$.

Step 2 Selection of initiating events: Define $\alpha \text{-} j_0$.

  Step 2.1: (*) (**) Identify first transition $\alpha \text{-} j_0$. Different classifications are used in PSA (see for instance (Mackowiak et al. 1985)) and Accident Analysis (see for instance Chapter 15.1 of (Almaraz 1986)).

  Step 2.2: (*) Evaluate $X_{\alpha j_0}(\vec{x}_0)$. See for instance (Mackowiak et al. 1985).

Step 3 Dynamic response model: Obtain the dynamics $\vec{g}_j(t, \vec{x}_0)$.

  Step 3.1: (**) Select a simulation driver. Note that a transition is a change in the status vector $j = (j_1, j_2, ..., j_N)$ and the driver should be able to change component models accordingly. Such simulation tools are already available. See for instance (Meléndez 1992).

Step 3.2: (**) Select plant model and data. This selection may depend on the initiating transient to be considered. In particular, determine the spectrum of possible j vectors to be considered.

Step 4 Reliability model: Obtain $P_{j \rightarrow k}(\vec{u} \rightarrow \vec{x})$ and $X_{jk}(\vec{x}_s)$.

Step 4.1: Get $P_{j \rightarrow k}(\vec{u} \rightarrow \vec{x})$. In the case that dependence on $\vec{x}$ is not considered (*) data can be taken from reliability data bases used in PSA (Chapter 5 of (US NRC 1983)). For fully deterministic analysis $P_{j \rightarrow k}(\vec{u} \rightarrow \vec{x}) = 1$. In the general case, the simulation driver should be able to evaluate $P_{j \rightarrow k}(\vec{u} \rightarrow \vec{x})$ which is no essential difficulty.

Step 4.2: Get $X_{jk}(\vec{x}_s)$. They may be taken from the results of front line system fault trees evaluations (*) (Chapter 6 of (US NRC 1983)) if dependence on $\vec{x}$ is not considered. Recent research (Macwan et al. 1991) is devoted to estimate the j, k and $\vec{x}$ dependencies, specially for operator actions.

Step 5 Dynamic Event Tree generation:

Step 5.1: Select a DET generating tool.

Select the technique to resolve eqs. (13), (14).

Select the approach for branching and expansion rules. Note that they may be strongly influenced by the DBEP selected for study, as for instance, bias techniques (Marseguerra, Zio 1992) toward top sequences.

Step 5.2: Obtain the DET by feeding the tool with the data obtained in the previous steps, and make parametric studies for different initial states.

Step 6 Evaluation of the acceptance criteria: Represent the results in the DBEP planes by plotting pairs $\left(\pi(\vec{x}, t), D_j(\vec{x}, t)\right)$ in the different regions and identifying potential top sequences. Verify that the initial states are safe states.

This method has the advantage that it may be made cumulative, as the probability that two different sequences have the same process vector is extremely small. Thus, whatever the completeness of the sequence and expansion rules is, partial results are useful. This is of particular interest to ensure feasibility, together with the considerations of the last paragraph of Section 1.

4.5 Selection of Tools

ISA requires the solution of the mostly deterministic dynamic event tree equations (i.e. eqs. (13) and (14), with the conditions of eq. (15)), with an appropriate set of sequence and expansion rules (Siu 1992) to treat the stochastic contributions. Present research is producing already able tools like DYLAM and similars like DETAM (Acosta, Siu 1991) and ADS ((Macwan et al. 1991), (Hsueh 1992)).

Monte Carlo techniques are also being developed (Devooght, Smidts 1992ab). All these methods have been so far designed with problem specific simulation of the dynamic paths of the process states and will only be appropriate for ISA if coupled with existing quick, flexible and powerful simulation tools in order to handle the complexities of control, limitation, trip and safeguard automatic/manual actions as well as the large number of physical phenomena involved in the process state evolution.

A suitable analysis tool, specifically designed for this purpose, is presently under development in order to gain this flexibility. The system will include a Dynamic Event Tree generator + simulation driver (DYLAM-TRETA-like module), as well as operator, control system and process states dynamic modules. Since the size and nature of the physical phenomena to be simulated will not be intrinsically limited, a modular structure and a suitable set of library modules will allow the selection of the level of detail. The driver will have capability to be coupled at each time step with problem-specific simulation tools that make feasible realistic descriptions.

References

Acosta C.G., Siu N. 1991 Dynamic Event Tree Analysis Method (DETAM), for accident sequence analysis, NUREG/CR-5608 draft version.

Aldemir T. 1991 Utilization of the cell to cell mapping technique to construct Markov failure models for process control systems, Proc. PS AM Mtg., Los Angeles, California, April.

Almaraz Nuclear Power Plant 1986, Final Safety Analysis Report.

Amendola A., Reina G. 1981 Event sequence and consequence spectrum: A methodology for probabilistic transient analysis, Nucl. Sc. Eng. 77, 297-315.

American National Standards Institute 1983 Nuclear safety criteria for the design of stationary pressurized water reactor plants, ANSI N 51.1.

Atomic Energy Commission 1972 Standard format and content of Safety Analysis Reports for Nuclear Power Plants. Regulatory Guide 1.70.

Barhucha-Reid A.T., 1960 Elements of the theory of Markov processes and their applications, McGraw-Hill, New York.

Cacciabue P.C., Cojazzi, G., Hollnagel E., Mancini S., 1992, Analysis and Modelling of Pilot-Airplane Interaction by an Integrated Simulation Approach, presented at 5th IFAC/IFIP/IFORS/IEA Symposium on Analysis, Design and Evaluation of Man-Machine Systems (MMS'92), The Hague, The Netherlands, june 9-11 1992.

Cojazzi G., Cacciabue P.C., Parisi P. 1992a, DYLAM-3, A Dynamic Methodology for Reliability Analysis and Consequences Evaluation in Industrial Plants, Theory and How to Use, ISEI/SER 2192/92 DRAFT.

Cojazzi G., Meléndez E., Izquierdo J.M., Sánchez M. 1992b The reliability and safety assessment of protection systems by the use of dynamic event trees. The DYLAM-TRETA package, Proc. XVIII annual meeting Spanish Nuclear Society, Puerto de Santa Maria, 28-30 October 1992.

Devooght J., Smidts C. 1992a Probabilistic Reactor Dynamics-I: The theory of continuous event trees, Nuc. Sci. and Eng., 111, 229-240.

Devooght J., Smidts C. 1992b Probabilistic Reactor Dynamics-II: The theory of continuous event trees, Nuc. Sci. and Eng., 111, 241-256.

Devooght J., Smidts C. 1992c Probabilistic dynamics: the mathematical and computing problems ahead, in T. Aldemir et al. (eds.) Reliability and Safety Assessment of Dynamic Process Systems. NATO ASI Series F, Vol. 120. Berlin. Springer-Verlag (this volume).

Devooght J., Izquierdo J.M. Relationships between probabilistic dynamics, dynamic event trees and classical event trees. To be published.

Izquierdo J.M., Villadóniga J.I. 1978 Basic philosophy of the nuclear regulation, ENS/ANS Brussels meeting.

Izquierdo J.M. et al. 1987 TRETA: a general simulation program with application to transients in Nuclear Power Plants, Revista de la Sociedad Nuclear Española, October, in spanish.

Izquierdo J.M., Sánchez-Perea M., Cacciabue P.C. 1988 Dynamic Reliability as a tool for the assessment of protection systems software analysis, presented at the NUCSAFE ENS/ANS conference, Avignon, France, October 1988.

Izquierdo J.M., Sánchez-Perea M. 1990 DYLAM-TRETA: An approach to protection systems software analysis, in Advanced Systems Reliability Modelling, Proc. Ispra course held at ETSI Navales, Madrid, Spain, September 1988.

Izquierdo J.M., Van Hoënacker L. 1992. Experience from thermalhydraulic plant calculations. Merits and limits. CSNI Specialist Meeting on Transient 2-Phase Flow. Aix-en-Provence (France), April.

Kafka P. 1992 Approximations of the Dynamic System Behaviour Within the Process of PSA, in T. Aldemir et al. (eds.) Reliability and Safety Assessment of Dynamic Process Systems. NATO ASI Series F, Vol. 120. Berlin. Springer-Verlag (this volume).

Mackowiak D.P., Gentillon C.D., Smith K.L. 1985 Development of transient initiating event frequencies for use in Probabilistic Risk Assessments, NUREG/CR-3862.

Macwan A.P., Hsueh K.S., Mosleh A. 1991 An approach to modelling operator behaviour in integrated dynamic accident sequence analysis, IAEA-SM-321/4

Marseguerra M., Zio E. 1992 Non linear Monte Carlo reliability analysis with biasing towards top event, accepted for publication in Rel. Eng. & System Safety.

Meléndez E. 1992 DYLAM-TRETA linkage, ISEI/SER technical note 2296/92.

Speis T.P. 1985 Licensing and Safety Analysis Background, in Proceedings of the International Center for Heat and Mass Transfer, Hemisphere Publishing Corporation.

Siu N. 1992 Risk Assessment for Dynamic Systems: An Overview, Submitted for publication to Rel. Eng. & System safety.

US Nuclear Regulatory Commission 1983 PRA procedures guide, NUREG/CR-2300

Part 3

Human Reliability and Dynamic System Analysis

Human Reliability and Dynamic System Analysis: An Overview

Ali Mosleh

University of Maryland, Materials and Nuclear Engineering Department
College Park, MD 20714-2115 USA

Abstract. Human interactions often play a critical role in determining the reliability of systems. There is an increasing recognition that accurate assessment of the reliability of dynamic systems requires integrated human and system model. This paper gives an overview of the issues and introduces several related works that address them.

Keywords. Human reliability, integrated reliability models

In a large number of cases, what contributes the most to the dynamics of a system is the human interaction. In the broadest sense, the human element is an integral part of any engineered system considering the life cycle of design, manufacturing, installation and operation. From a reliability assessment point of view, human interaction in all of these phases are important. For instance, human errors during design, not detected and corrected during reliability demonstration tests, could surface during operation as latent causes of malfunction. Similar latent failures could be introduced, for example, during operation or maintenance of the system. While these types of interactions and errors could be viewed as time dependent man-machine interactions in the time scale of the life of a system, what is of interest here are interactions with shorter time constants as part of system/process control during normal operation or in response to abnormal and accident conditions.

Factoring this type of human interactions in the reliability or risk assessment of a system involves: 1) identification of the type of interaction, 2) incorporation of the interaction into the reliability model, and, 3) quantification of the probabilities or other parameters of the model that measure the impact of the human element on the reliability or risk level.

Clearly all three aspects are interrelated and a comprehensive methodology should address them at an adequate level of detail. Nevertheless the degree to which these aspects have been treated varies among methodologies.

Over the past two decades, particularly in the past ten years, considerable effort has been devoted to addressing each of the various aspects of reliability implications of man-machine interactions. Much of the work with emphasis on quantifying human error probabilities has emerged in relation to probabilistic risk assessment of nuclear power plants where a number of major accidents have been attributed to human failure. Examples of these human reliability analysis (HRA) methodologies the Technique for Human Error Rate Prediction (THERP) [1], and SLIM-MAUD [2].

The extent of man-machine interactions during normal as well as abnormal operating conditions depends on the degree of automation in control and accident response of the system, and the flexibility provided in terms on manual override of automatic controls. Furthermore, the interaction points and types may be guided by procedures. Nevertheless, irrespective of the what the designer has envisioned as being the spectrum of operator-system interactions, what concerns a reliability analyst is whether the operator can change the system state through intentional or unintentional acts. With the spectrum of possible interactions identified, the cause(s), likelihoods and impacts of the actions can be investigated.

Currently the way man-machine interactions are incorporated in the risk and reliability models (e.g., fault trees, event trees and Markov models) is based on the same underlying philosophy of representing the impact of the action in form of change in the system state. In other words, the reliability models are almost universally hardware-oriented. This means that the operator is modeled only by the manifestation of his/her action, leaving the modeling of the causes and mechanisms of the action as well as the computation of its likelihood to a separate and often disjointed, effort. This has certainly been the case in the treatment of operator actions in models of nuclear power plant accident sequences where scenarios are formed by the combinations of hardware success and failures and success or failure of operators to initiate or terminate functions (as guided by procedures).

A major pitfall of the segregation of the model of the man from model of the machine is in the tendency that it creates to oversimplify the interaction and the impact of its dynamics. This may not be an inherent limitation of the "modularization" of the models but it is a widespread phenomenon in practice.

In recent years some researchers in both areas of systems reliability and human reliability have come to the conclusion that a more realistic approach to modeling man-machine interaction in dynamic systems is an integrated one in which the system and the operator are considered as inseparable elements of same reliability model. On the system hardware reliability analysis side the impact of operator actions on the time evolution and dynamic dependencies of system states during normal or abnormal operation are being viewed as important factors for accurate deterministic and probabilistic assessment of system behavior. In the HRA domain there is a widespread realization that going beyond the oversimplifications of the current human response models require understanding of the causes of human action. This, inevitably brings the consideration of the "context" of the interaction into the forefront of modeling activity. This is why there is an increasing interest in the so called integrated models, dynamic reliability approaches and cause-based and cognitive HRA models.

In the past few years, the efforts directed at developing the "next generation" of HRA models have gained significant momentum. Central to these efforts is a better qualitative understanding of the causes of human error. The search for cause from some researchers point of view should include the roots in factors such as characteristics of the organization behind the man and the machine while recognizing the importance of the immediate context of the interaction. This view is echoed in [3]. Work is also being directed at developing the outline and general characteristics of improved HRA models. A systems perspective is essential for limiting the scope and requirements of such models [4]. This is in part due to the concern that development of a very detailed, generic, cognitive model that covers a broad spectrum of human interactions may not be possible within the current state of the art in cognitive and behavioral sciences. In this respect some experts advocate moderation and balance in modeling. Simplification of the complex problem of human modeling and the use of simulation environment is seen as an effective approach to realize the next generation of HRA models [5].

References

1.   Swain, A.D. and Guttman, H.E.: Handbook of Human reliability analysis with emphasis on nuclear power plant applications. NUREG/CR-1278, U.S. Nuclear Regulatory Commission (1983)

2.   Embrey, D.E. et al: "SLIM-MAUD: An approach to assessing human error probabilities using structured expert judgement. NUREG/CR-3518, U.S. Nuclear Regulatory Commission (1984)

3   Wreathall, J.: Human errors in dynamic process Systems. These proceedings.

4   Parry, G.: Critique of current practice in the treatment of human interactions in probabilistic safety assessment. These proceedings.

5.   Hollnagel, E.: Simplification of complexity: the use of simulation to analyze the reliability of cognition. These proceedings.

Critique of Current Practice in the Treatment of Human Interactions in Probabilistic Safety Assessments

Gareth W. Parry
NUS, 910 Clopper Road
Gaithersburg, MD 20878, USA

Abstract. The current human reliability approaches adopted in nuclear power plant probabilistic safety assessments are deficient in some important ways. The treatment of dependency between the human interaction events of the logic models is of particular concern, and errors of commission have not been addressed in a comprehensive way. Arguments are presented that an appropriate way to resolve these problems is to develop Human Reliability Analysis models that have embedded in them the ability to address causes of error, and recovery from those errors based on feedback from the plant. A potential role for dynamic PSA models is identified.

Keywords. human reliability, dependency, errors of commission, human error probabilities, performance influencing factors.

1 Introduction

The importance of including a consideration of the human element in Probabilistic Safety Assessments (PSAs) of nuclear power stations has long been appreciated, and, as the discipline of PSA has matured, increasing attention has been paid to modeling the impact of plant personnel interactions with the plant equipment . There are two important considerations for the modeling of human interactions in the context of a PSA. The first is the representation of those interactions in the logic structure of the PSA model, and the second, the quantification of the impact of those human interactions. A successful PSA involves close cooperation and iteration between the systems analysts, who have the primary responsibility for developing the structure of the logic models, and the human reliability analysts, who have the primary responsibility for quantifying the human error probabilities.

The general characteristics of the treatment of human interactions in PSAs are summarized in the next section. This is followed by a critique of the state of the art. In the final section, some suggestions for improving the methodology are presented, and a potential role for dynamic PSA techniques is discussed.

2 The Treatment of Human Interactions in PSA

A PSA model is constructed to identify and estimate the frequencies of the scenarios that lead to core damage or other undesirable plant state. In the PSA context, a core damage scenario is defined by specifying an initiating event, and a series of equipment failures and/or operator errors, which, together, lead to failure to cool the core. There is a considerable variety of approaches to constructing PSA models, although all have a common feature. They all involve the construction of event trees, which are inductive logic diagrams, as the fundamental framework of the model. The major discriminant between different PSA approaches lies in the amount of detail incorporated into the event tree structure. This ranges from the relatively small functional event tree, through the larger systemic event tree, to the support state event tree model. The accident sequences defined in these event trees are increasingly more precisely defined in terms of equipment unavailability. Nevertheless, the sequences, even in the most detailed event trees, are defined in terms of fairly high level supercomponents, which could correspond to a complete system or the individual trains of a system. Reliability models, such as fault trees or reliability block diagrams, are used to increase the definition of sequences down to the level of component unavailability states. Incorporated in this set of logic models are events that represent the impact of plant-personnel interactions. These events may be included as event tree headings, in functional fault trees or function equations (sometimes called event tree function top logic), or in system fault trees.

There are three classes of human interactions that are, in principle, addressed in the models. There are interactions that occur prior to the initiating event, and are related to maintenance, testing, and calibration activities. While these activities are essentially positive, if errors are made, they can lead to an unrevealed unavailability of equipment that might be required following an initiating event. It is these errors that are included by basic events that represent such failure modes of equipment as, "valve in wrong configuration", or "sensor miscalibrated". The human errors that contribute to these events are often called latent errors, since they have no immediate effect, but are only revealed when the affected equipment is demanded.

There are other interactions that are related to the normal plant operations, and also to some testing activities, where errors may inadvertently lead to plant trips. These are very rarely explicitly included in PSA logic models; it is usually argued that their impact is already included in the initiating event frequencies.

The third class of interactions includes those taken by the control room operators in response to certain combinations of equipment failures which lead them to take specific actions as directed by the Emergency Operating Procedures (EOPs), Abnormal Operating Procedures (AOPs), and Functional Restoration Procedures (FRPs). The primary impact of these interactions is again clearly positive. However, the possibility that the actions are not performed successfully is included by incorporating, in the model, events that represent failure of these actions. In addition, interactions that represent the possibility of more innovative actions, which are not prescribed by current procedures, are also considered. These are often called recovery actions. Again, the possibility of failure of a recovery action is represented by a basic event of the PSA model.

The modeling of human interactions is completed by estimating the probabilities of the events that represent failure resulting from human error.

3 A Critique of Current PSA Approaches

The two major recurring criticisms of the treatment of human interactions in PSA are: (a) PSAs do not generally address errors of commission, and (b) there is no universally accepted, robust method of quantifying human error probabilities. These two issues are elaborated on in turn below.

3.1 Errors of Commission

As discussed previously, a PSA logic model identifies undesirable plant states in terms of scenarios (essentially accident sequence minimal cutsets) which consist of an initiating event, and combinations of events that represent the unavailability of equipment to perform its required mission, whether as a result of failure or from other causes. Human interactions are included through events that represent causes of unavailability of specific pieces of equipment (e.g., valve left in wrong configuration), or that represent failure to perform a required action (e.g., failure to initiate Residual Heat Removal). Most recent PSAs have invested substantial effort in identifying the opportunities for failures resulting from human error.

As alluded to above, interactions that are associated with operating crew responses following an initiating event are usually introduced through a binary, success/non-success, logic; failures are modeled as non-response, i.e., as errors of omission. Other failure modes, such as specific inappropriate acts resulting from human error, which cause transitions between different accident scenarios, are not usually explicitly identified, and the impact of their consequences is therefore missing from the model. Because, ultimately, PSA scenarios can be interpreted in terms of the availability/unavailability or configuration of the plant equipment, it can be argued that most possible scenarios are already imbedded in the structure of the logic model. (One possible group of scenarios not necessarily included is that involving spurious actuation of systems; spurious, non-operator induced actuation is generally considered a low probability event and may have been screened out of the model.) However, as the above discussion implies, it has to be recognized that their frequencies are not estimated correctly unless the probabilities of transition between scenarios due to inappropriate actions have been factored in. These inappropriate actions are the errors of commission of most interest to the PSA analyst.

For the pre-initiating event, and initiating event related human interactions, many errors that lead to undesirable states are indeed included in the quantification of the model. For example, errors, or inappropriate acts that lead to an initiating event, or that lead to the unrevealed (latent) unavailability of equipment may be represented in the databases from which the initiating event frequencies or equipment unavailabilities respectively, are estimated. However, to understand to what extent this is the case would require investigating the detailed contents of the databases, and the way the event frequencies or probabilities are defined and estimated. From a PSA perspective, the primary concern with errors of commission is likely to be associated with the evaluation of those rare initiating events which are treated by constructing logic models (e.g., loss of component cooling water system) rather than by the use of historical data to estimate frequencies, or those cases where there is a potential common link between the initiating event and the required response. This may be relatively more important for the modeling of the shutdown phase, than for the full power phase.

3.2 Models for Quantifying Human Error Probabilities

The quantification of the human error probabilities associated with the latent, or pre-initiating events is generally performed using THERP[1], which, in this domain at least, has a large degree of acceptance. However, there is little in the way of consensus in the quantification of error probabilities associated with post-initiating event, control room guided responses.

It is not the purpose of this paper to provide a comparative criticism of the various methods proposed. This has been done in several reviews in recent years[2]. It is noted, however, that there are several different approaches, probably as many (if not more than) the number of practitioners. There are areas of similarity among these methods, and areas of disagreement. Most approaches seem to be based on the premise that the human error probabilities can be expressed as a function of influencing factors, sometimes called performance shaping factors. It is in the approach to determining what are the key influencing factors, and characterizing the impact of those influences, that the differences lie. Some approaches explicitly address several performance shaping factors[3,4], while others pick out one or two key factors[5,6]. It seems to be generally true, however, that there is relatively little in the way of a firm theoretical basis for any of these models, although there may be some empirical support for their applicability. However, even in these cases, application to PSAs requires a considerable amount of extrapolation.

The models, however, do have one major feature in common. They all attempt to increase the human error probabilities from scenario to scenario, as the scenarios become more demanding. In this way, an attempt is made to capture the scenario dependence of the error probabilities in a relative way.

Another issue associated with quantification that is a major source of concern is that of probabilistic dependency. In many PSA models, several human interactions basic events may occur in the same accident sequence cutset, or scenario description. These events may not be independent; the probability of the second, third, etc., event may be a function of whether the first failure event occurs. This dependence generally implies some shared cause, or influence factors. Current models, because they do not address causes to any major degree, have difficulty in addressing this dependence quantitatively[7]. A consequence of ignoring this dependency is the possibility of inadvertently suppressing the importance of human responses in certain scenarios. PSA analysts are, however, increasingly aware of the possibility of such dependencies, and while their quantification may not be justified on the basis of a theoretical model, the scenarios where such dependency exists can be identified and steps taken to ensure that they are not screened out of the analysis too early[8].

4 Discussion of Potential Improvements

While current PSAs have several immediate uses, their usefulness as decision making tools, could be considerably improved by enhancing the analysis of human interactions. It is clear that the principal areas of improvement should address both the issue of errors of commission, and the quantification models themselves, particularly with regard to the treatment of dependency. In this section, some suggestions are made on what such a Human Reliability Analysis (HRA) methodology might look like. These suggestions represent partial results of an ongoing Electric Power Research Institute (EPRI) sponsored project to define a plan for developing such a methodology.

4.1 Definition of Problem

The first step is to develop a clear functional definition of what we mean by a human error. As a starting point, it can be stated that an error can be recognized as having been made when there is evidence of a level of performance that is inadequate with respect to an applied criterion. This is fairly standard "definition" but it is too imprecise as a working definition in the context of system modelling, since it allows an error to be defined at any level of human interaction description. Fortunately, in the context of creating analytical models of complex technological systems the definition can be made more precise as follows, using the natural scale provided by the degree of detail provided in the model.

As discussed in Section 2, the modeling techniques used for safety and reliability studies evolve around developing logic structures whose basic elements can be identified with various states of the component parts of the system. Thus, inadequate performance can be identified with inappropriate transitions between states of components, or in the case of non-response, the non-existence of transition that would be required to help bring the plant to a safe shutdown. In this way, the human errors of interest can be identified in terms of their effect rather precisely, either as failure mechanisms of specific equipment or functions, or as inappropriate actuation mechanisms which lead to failures of functions. Stated another way then, the errors in which we are interested are expressed as failure modes of components, systems, or functions.

The next question to be addressed is, "what is it about these errors that we wish to study and why?" In the context of PSAs that are performed to provide some quantitative measure of risk, the goal could be simply stated as the estimation of human error probabilities associated with the "component" failure modes discussed above. There are those (e.g., [9]) who claim that this is a hopeless task, and that human behavior is so conditioned by internal and external influences that any probability has to have such large uncertainties as to be essentially meaningless. However, the intent in PSAs is not to capture individual differences between operating crews, as indeed it is not to try to capture differences between the reliability characteristics of individual members of a population of redundant components. Instead, it is the time averaged and population averaged reliability characteristics that are explored. In addition, PSAs model human interactions in a context specific way in that the set of hardware failures and successes and failures up to a certain point in the scenario development determine the requirements for subsequent human actions, and dictate some of the commonly identified external performance-shaping factors.

It is accepted as a starting point that human errors occur and that it is appropriate to deal with their occurrence in the PSA context as random processes. This may, but does not necessarily imply that humans behave in a random manner. It may also reflect the fact that PSA scenarios are by their nature imprecise in defining the conditions that may influence human behavior. This imprecision leads to a probabilistic description. As a limiting case, it is also accepted that certain circumstances guarantee failure, i.e., the human error probability (HEP) equals 1.

4.2 General Characteristics of an HRA Methodology

Ultimately, it is desirable to have a predictive theory of errors as they are manifested in terms of the impact on the system. The theory should be predictive not in the sense of identifying <u>when</u> an error will be made, but in the sense of having the capability of recognizing the signature of particularly error prone situations or scenarios and of providing a means of assessing the relative likelihood of the different ways in which errors may be expressed in these different scenarios.   The interface with a PSA systems model is in the definition of the <u>error expression</u>, i.e., how the error is manifested in the context of the system. There are several categories that could be considered in PSAs:

    a. inadequate response, which can be divided into
- no response
- slow response
- premature response
    b. alternate response, of which there may be many possibilities.

In general, as discussed earlier, only the first two are normally modelled in current PSAs. The definition of an error expression is completed by specifying which specific equipment or function is affected and how. An error expression is, therefore, a failure mode of a piece of equipment or a function. As will be seen in the following, this 'error' in the time-frame of the operator interaction with the system is not necessarily an isolated, point event, but more generally, corresponds to failure of a process.

For each error expression, the model should be capable of addressing the different <u>error modes,</u> corresponding to proximate causes of the error expression, e.g., chose incorrect procedure, failed to see alarm, etc. An error mode is a quasi-phenomenological description of a failure mechanism as it is manifested in the interfaces between the operators and the system and/or information base. The error modes give some explanation of how the failure (or error expression) occurred and what the associated error was, but not <u>why</u> the error was made.   Typically errors may occur in the operator information interface, in the formulation of a response, or in its execution.

The next level of detail involves identifying for each mode, <u>error mechanisms or error causes</u> which explain why the errors occur. For PSA purposes this explanation may be best characterized by specifying scenario specific factors we will call <u>performing influencing factors</u> (PIFs), although since our primary interest is errors, they might be renamed <u>error influencing factors</u> (EIFs) for those factors that have a negative influence, and <u>error compensating factors</u> (ECFs) for those that tend to decrease the likelihood of errors. Understanding how and why a particular PIF, or combination of PIFs impacts the likelihood of success or failure, in a given task will clearly involve understanding the operator's cognitive processes.   Errors in a PSA context, are failures of a process, and that process, in most cases, allows recovery from an initial error partly because the inertia of the system does not lead immediately to failure of vital equipment, or to an irreversible plant state, and therefore the operators have the possibility of receiving feedback from the plant which allows them the opportunity to recognize and recover from errors. To complete the modelling of human interactions, and to allow the estimation of the probabilities of error expressions it is therefore important to incorporate recovery mechanisms. Since different errors may result in different system responses, the feedback to the operators will also be different, and this will impact the potential for recovery. The potential for recovery is also a function of the error cause. For example, errors of intention are generally felt to be less likely to be recovered than are simple slips.

The above considerations suggest that an HRA model should recognize that the different error causing mechanisms may be affected by different PIFs and that they have different recovery potential. One model having some of these characteristics has been reported in [10]. In a model based on this concept, an error expression is the logical sum of contributions from several error modes, which in turn can result from several error mechanisms modified by appropriate recovery mechanisms.

4.3 Some Thoughts on Definitions of Appropriate PIFs

The challenge for developing a model that is simple enough to meet PSA needs is to identify appropriate PIFs to characterize PSA scenarios. This section offers some suggestions on the nature of these PIFs.

Based on Reason's[11] hypothesis that the dominant cognitive primitives for information retrieval are similarity matching and frequency gambling, it is reasonable to propose that scenarios whose signatures are similar to those of other more commonly experienced or expected scenarios are candidates for misdiagnosis. Other scenarios, characterized by information overload or excess stress, may lead to resorting to "familiar" behavior. A common factor is that there is some similarity between the signature of the actual scenario, and that assumed by the operator. This suggests that one way to define a PIF might be as some measure of similarity between the signature of the actual scenario, and that perceived by the operator which led to the error.

Another type of error producing mechanism that results in a delayed response, is a result of a conflict between the benefits and adverse consequences of performing the action. This again may be characterized by a comparative PIF.

Furthermore, the recovery process depends on the signature of the state into which the plant/system is directed as a result of the initial error. The potential for recovery might be a function of the difference between the signature of the incorrect state and that of the correct state, or that of the state the operator thinks the plant should be in. Thus the PIFs that characterize the recovery process perhaps ought to reflect the quality of the feedback, measured as a function of the difference in signatures mentioned above.

Based on these observations it may be fruitful to explore models that use some comparative measures of independent observables as PIFs. Certainly, a model based only on the PIFs related to the scenario that cues the action will not be adequate.

4.4 A Role for Dynamic PSA

It is clear that the evaluation of a human error probability should include the complete set of sequences of operator/plant interactions that can lead to an error expression. This is a highly dynamic process, and perhaps the only way it can really be explored is by using the simulation approaches of dynamic PSAs. These approaches enable an analyst to investigate the interplay between the PIFs, the role of the recovery process, and in particular any mechanisms for dependency between causes of initial errors, and failures in the recovery process. However, it is not obvious that, once the processes have been well understood, that an HRA model for use in the standard PSA approach cannot be formulated as an adequate approximation.

Acknowledgments

The author has benefitted from discussions with D. H. Worledge of EPRI, and A. Mosleh, K. S. Hsueh, A. Macwan and S. H. Shen of the University of Maryland, particularly on the topics addressed in Section 4. The support of EPRI is appreciated.

References

1. Swain, A. D., and Guttman, H. E.: Handbook of Human Reliability Analysis with Emphasis on Nuclear Power Plant Applications. NUREG/CR-1278, August 1983.
2. See, for example, Swain, A. D.: Comparative Evaluations of Methods for Human Reliability Analysis, GRS mbH, ISBN 3-923875-21-5, Garching, Germany, April 1989.
3. Williams, J. C.: A Data Based Method for Assessing and Reducing Human Error to Improve Operational Performance, IEEE, 1988.
4. Embrey, D. E., et al: SLIM-MAUD: An Approach to Assessing Human Error Probabilities Using Structured Expert Judgement, NUREG/CR-3518, July 1984.
5. Spurgin, A. J., et al: Operator Reliability Experiments Using Power Plant Simulators, 3 Volumes. EPRI NP-6937, July 1990.
6. Dougherty, E. M., Fragola, J. R.: Human Reliability Analysis: An Engineering Approach with Nuclear Power Plant Operations. John Wiley, New York, 1988.
7. Parry, G. W., and Lydell, B.O.Y.: HRA and the Modeling of Human Interactions, in Probabilistic Safety Assessment and Management, Ed. Apostalakis, G. A., Elsevier, 1991.

8.   Wakefield, D. J., Parry, G. W., Hannaman, W. G., Spurgin, A. J. and Moieni, P.:   Systematic Human Action Reliability Procedure (SHARP) Enhancement Project, SHARP 1 Methodology Report, EPRI, (to be published).

9.   Moray, N.: Technical Note: Dougherty's Dilemma and the One-Sidedness of Human Reliability Analysis, Reliability Engineering and System 29 (1990), pp. 337-344.

10.   Parry, G. W., et al: An Approach to the Analysis of Operating Crew Responses Using Simulator Exercises for Use in PSAs in Proceedings of the OECD/BMU Workshop on Special Issues of Level 1 PSA, GRS-86, July 1991.

11.   Reason, J.: Human Error. Cambridge University Press, 1990.

Simplification Of Complexity:
The Use Of Simulation To Analyse The Reliability Of Cognition

Erik Hollnagel

Human Reliability Associates, Ltd., 1 School House, Higher Lane, Dalton, Parbold, Lancs. WN 8RP, UK

Abstract: In order to be adequate, safety and reliability analyses of man-machine interaction must account for the dynamics of the interaction. There is therefore a need for models that can describe the dynamics of human cognition and performance. Most existing models are developed for static analyses and are too restrained in their assumptions to serve as a basis for simulations. The paper describes a contextual control model of cognition, which focuses on how the choice of actions is determined by how the current situation develops. This model has been used as part of a joint man-machine simulation tool designed to evaluate the effect of human actions on system performance.

Keywords: human reliability, simulation, context, control, cognitive model, human error

1   Analysis = Simplification

Safety and risk analyses are one way of compensating for the unavoidable incompleteness of design and implementation. The basic approach is to identify the various ways in which a malfunction may occur, the ways in which it may develop, and the ways in which it may be prevented, eliminated, or contained. In order to accomplish this it is, however, necessary to reduce the complexity of the system being described to a level which is amenable to analytical methods.

The aim of an analysis is to explain something by concepts which are more fundamental than those used to describe the phenomenon as it appears (the manifestation, the phenomenal level). It follows that the phenomenon can only be described to the extent that it can be brought to match the terms and concepts that are used by the analysis. A reduction of the description is necessary to make an analysis possible. The art is to make the right reduction - and to recognise that a reduction has been made.

1.1 Point-To-Point Analyses

The simplifications that are imposed by an analysis are easily illustrated by considering the main approaches. Common to all analyses are the concepts of a cause (a source) and a consequence (a target). The purpose of the analyses is to find the possible links between causes and consequences, in order to devise ways in which the activation of these links can be avoided. Four situations can be considered:

o   Single cause, single consequence: This investigates the specific relations between a single cause and a single consequence. Examples are root cause analysis and consequence propagation techniques.

o   Single cause, multiple consequences: This investigates the possible outcomes of a single cause. Typical techniques are Failure Mode, Effects and Criticality Analysis or Sneak Analysis.

o   Multiple causes, single consequence: This investigates the possible causes for a specific consequence. Examples are Fault Tree Analysis and Common Mode Failure analysis.

o   Multiple causes, multiple consequences: This investigates the relations between multiple causes and multiple consequences. Examples are Hazard and Operability Analysis, Cause-Consequence Analysis, and Action Error Mode Analysis [1].

These analyses can all be characterized as **point-to-point analyses** because they are based on a limited number of pre-defined analysis points. The advantage of point-to-point analyses is that they focus on relevant subsets of the domain, e.g., safety critical or mission critical events. The limitation of point-to-point analyses is that the quality of the outcome is restricted by the number of points (causes and consequences) that are considered. The difference between field studies and analytical methods is that the former address a reasonably large subset of the actual events without imposing too many conditions on the description, while the latter only consider a formalised representation of the main features. Field studies are basically a question of discovery; analyses are a question of systematically uncovering the fundamental principles that are (assumed to be) the causes of observed regularities - or rather, in the case of PRA, irregularities.

2   The Differences Between Humans And Machines

The fundamental principle of engineering methods is decomposition, which also is at the heart of all attempts to capture human performance by means of event trees [2, 3]. Unfortunately, decomposition is not acceptable as a basis for describing and analysing human performance. When it comes to human performance, and in particularly human cognition, we only have vague ideas about their principle of function, we know nothing of their design, we have few models available, we cannot delimit their range of applications, we are uncertain about the failure modes, and there are no fixed upper and lower bounds for their capacity. Consequently the human can neither be modeled nor analysed in the same way as a component or an aggregation of components. Human performance is a result of a complex interaction with the context which lies beyond the range of simple deterministic descriptions. The solution is to recognise the fundamental differences between humans and machines, and develop a basis for reliability analysis which takes these differences into account.

Considering the human operator in machine-like terms is only warranted when the human is trained to work as a machine (as in Scientific Management or highly regulated areas of operation), or in the case of highly skilled performance that can be carried out automatically - e.g. riding a bicycle. In all other cases human performance is embellished by human cognition - thinking rather than doing - or the human inclination to try to do things in a different way, either out of boredom, to be more efficient, to avoid unnecessary work, etc. Having recognised the importance of human cognition, it soon proves to be an obstacle for human reliability analysis. There are a number of reasons for that.

o   Firstly, human cognition is **covert rather than overt**; it cannot be observed (except by introspection which is a process with limited validity).

o   Secondly, and partly as a consequence of that, there is a **lack of good theories of cognition** even though the recent wave of cognitive psychology goes back to the 1960s. Most of the theories that we have are typically descriptions of characteristic or dominant modes of performance - notably decision making, problem solving, and diagnosis. They are theories of specific tasks rather than of performance in general.

o   Thirdly, in the cases where we do have models of theories of cognition, they are **impossible to validate**. There is little possibility of proving them correct. The best one can do is trying to disprove them [7] and maintain them until this has happened.

Altogether this means that the basis for analysing human cognition, hence human reliability and the reliability of cognition, is pretty flimsy. While it is clearly necessary to simplify the description of human performance in order to analyse it, the question remains how this simplification shall be brought about. In the following I will consider two principally different ways of doing that: procedural prototypes and contextual cognition.

3 Models Of Cognition

In order to analyse the reliability of cognition it is necessary to have a clear notion of human cognition and the nature of work. In other words, it is necessary to have an adequate model of the human - an operator model or a metaphor for human information processing. Human performance can be accounted for in several different ways, ranging from the Stimulus-Organism-Response (S-O-R) paradigm and view of the human as an Information Procesing System (IPS) to the cognitive viewpoint [4], exemplified and expanded by cognitive systems engineering [5, 6]. of which the most prominent are described below. The main difference from the S-O-R and the IPS metaphors is that cognition is viewed as active rather than re-active. The assumptions are that behaviour is not simply a function of input and system (mental) states and that the complexity of the inner "mechanisms" is so high that it **cannot** be captured adequately by a single theory. The focus of the cognitive viewpoint is therefore on **overall performance** as it appears rather than on the **internal mechanisms** of performance.

3.1 Procedural Prototypes and Contextual Cognition

In models of cognition a distinction can be made between those which emphasize the sequential nature of cognition and those which view cognition as being determined by the context.

o   A **procedural prototype model** implicitly expresses the view that there exists a characteristic or pre-defined sequence of (elementary) actions which represents a more natural way of doing things than others - or even that a certain sequence or ordering is to be preferred. According to this view the expected next action can, at any given time, be found by referring to the ordering of actions implied by the prototype.

o   A **contextual control model** implies that actions are determined by the context rather than by an inherent relation between them. It is therefore not possible *a priori* to describe procedural prototypes or "natural" relations between actions; the choice of the next action at any given point in time is determined by the current context (conditions). A contextual control model therefore concentrates on how the control of the choice of next action takes place rather than deliberating on whether certain sequences are more proper or likely than others.

The difference between the two views lies not in the phenomena they address but in the way they do it, i.e., the concepts and relations that are used for descriptions and explanations. The procedural prototype models reify the recurrent patterns as templates or standards for action. The contextual control models instead look to the factors that produce the recurrent patterns. In the first case much effort is therefore spent on determining the extent to which actual performance complies with "prescribed" performance, and how much performance can be shaped or forced to conform to the normative description, e.g. through design of interfaces, working conditions, procedures, training, organizations, etc. In the second case the investigations are shifted towards the way in which actions are actively selected, as part of the person's coping with the current situation.

3.2 The Separation Between Competence and Control

A contextual control model of cognition has two parts: a **model of competence** and a **model of control**. A person has a large number of possible actions or functions available. In the model of competence the possible actions provide a set to choose from on different levels of aggregation, i.e., a mixture of simple and composite actions. Common to them is only that they are ready to use in a given context. Thus for each given domain or set of tasks it is possible to define a small set of frequently recurring actions. The selection from these possible actions is determined by the existing needs and constraints rather than by normative characteristics of the elementary or component actions (the control). Although there may be frequently recurring patterns or configurations of actions, there is no pre-defined organization or prototype. It is therefore necessary to provide a reasonable account of how this can be the case, without making any unnecessary assumptions about human cognition.

The control function can be defined and described in a number of ways. In the contextual control model the pre-conditions, which are part of how actions are described, constitute the basis for tactical or strategic planning of actions. It is because we know that there are certain pre-conditions and use them as a basis for planning, that a certain sequentiality or ordering appears: the resulting sequentiality is brought about by explicit control rather than by the inherent characteristics of the elementary actions. If, therefore, a person is inexperienced or confused, improper planning may ensue and actions may be carried out in a counterproductive way. That is ample evidence that sequentiality is not an inherent feature of the elementary events or the model, but rather comes from the way control is exercised. A more detailed description of this model is given in [8].

4 A Simulation-Based Analysis Tool: The System Response Generator

The difference between procedural prototypes and contextual control becomes important when a simulation-based approach is used to analyse the reliability of cognition. The simulation is used as a basis for analysing the dynamics of the interaction between the user and the process. The emphasis must therefore be on developing a model which in a realistic way exhibits the requisite variety of the phenomenon, rather than on a model which is faithful to a specific theory.

A good example of a simulation-based analysis tool is found in the System Response Generator (SRG) [9]. The overall purpose of the SRG project is to develop and implement a software tool which can be used to analyse the influence of human decision making and action on the way in which incidents in complex systems evolve, in particular to (1) **identify potential problem areas**, i.e., the aspects of the task and the man-machine interaction where problems are likely to occur, and (2) **evaluate the effects of specific modifications to the system** (e.g. of procedures, information presentation, control options, etc.).

The concern of the SRG is with the MMI, specifically the way in which misunderstandings and incorrectly executed human actions can change how an accident evolves. Based on the DYLAM approach [10, 11], the SRG uses a consistent method for dynamically simulating both the technical process and the human operator. The operator modelling draws on state-of-the-art solutions in cognitive modelling and knowledge representation. The process model may use a mixture of modelling techniques to achieve the proper trade-off between efficiency and accuracy required by the SRG.

4.1 Operator Modelling in the System Response Generator

Process modelling in the SRG is achieved by means of available process simulators for the application in question. The SRG provides a well-defined interface protocol for establishing the interaction between the SRG and the process simulator, as well as a method to define the additional process knowledge which is needed for a scenario to run. The development of a specific process simulator is, however, outside the scope of the SRG.

The purpose of operator modelling in the SRG is to emulate the response of an operator under specific conditions. The range of responses (outputs) from the model must therefore go from correct response to incorrect or erroneous responses. To make it manageable for the SRG it is assumed that performance can be modelled by three different levels as follows:

o   On **Level-0** the performance is flawless and correct. This corresponds to an operator with correct knowledge, infinite capacity, and infinite attention. The input maps directly onto the output and the nominal or prescribed relations between input and output variables are reproduced in any suitable form. Level-0 serves the purpose of testing the consistency and correctness of e.g. procedure design and interface support.

o   On **Level-1** the performance is expressed by **quantitative modelling** in terms of continuous functions, e.g. a probability function. This corresponds to an operator who is influenced by the working conditions, and the independent variables will all have to do with external conditions, such as interface design, structure of procedures, work demands (leading to work load), noise, etc. Level-1 includes a large variety of models, such as probabilistic models, supervisory control models, filter models, differential equations, Markov models, etc. Their common feature is that they do not make any pretense of incorporating psychological (cognitive) concepts or functions, but achieve their purpose by means of continuous, mathematical functions (e.g. signal processing models).

o   On **Level-2** the performance can be expressed by **symbolic modelling** in terms of rules or heuristics. This corresponds to an operator who is influenced by the perceived conditions or internal states (cognitive states), e.g. assumptions, intentions, biases, preferences, habits, etc. The model is expressed in terms of symbol manipulation, typically as an information processing model.

Although the SRG will be provided with a default operator model the system is designed to work with an arbitrary operator model. In order to accomplish that the interface between the operator model and the SRG has been clearly defined, and the operator model functions as an external module, in the same was as the process model does.

4.2   The Contextual Control Model (COCOM)

In the SRG a specific version of a contextual control model is used; it is referred to as the Contextual Control Model or COCOM. Control is necessary to organise the actions within the person's time horizon. Control is partly based on planning what to do: this is influenced by the context as it is experienced by the person (e.g. the cognitive goals described by [12]), by knowledge or experience of dependencies between actions (pre-conditions, goals-means dependencies), and by expectations about how the situation is going to develop, in particular about which resources are and will be available to the person. In COCOM, a distinction is made between four different modes of control:

o   In **scrambled control** the choice of next action is completely unpredictable or haphazard. The corresponding situation is one in which the person is in a state of mind where there is no reflection or cognition involved, but only blind trial-and-error. This is typically the case when people act in panic, when thinking is paralyzed and there accordingly is little or no correspondence between the situation and the actions.

o   In **opportunistic control** the next action is chosen from the current context alone, based on the salient features rather than on more durable intentions or goals. The person does very little planning or anticipation, perhaps because the context is not clearly understood or because the situation is chaotic. In this type of situation the person will often be driven either by the perceptually dominant features of the interface or by those which due to experience or habit are the most frequently used, e.g. similarity matching or frequency gambling heuristics [13].

o   In **tactical control** the person's performance is based on some kind of planning, hence more or less follows a known procedure or rule. Although the time horizon transcends the needs of the present, planning is of limited scope or limited range and the needs taken into account may sometimes be ad hoc. If the plan is a frequently used one, performance corresponding to tactical control may seem as if it was based on a procedural prototype - corresponding to e.g. rule-based behaviour [14]. Yet the underlying basis is completely different. The regularity is due to the similarity of the context or performance conditions, rather than by the inherent "nature" of performance.

o   In **strategic control** the person is considering the global context, thus using a wider time horizon and looking ahead at higher level goals. The strategic level should provide a more efficient and robust performance, and thus be the ideal to strive for. The attainment of strategic control is obviously influenced by the knowledge and skills of the person, i.e., the level of competence; although all competence can be assumed to be available, the degree of accessibility may greatly vary between persons, hence influence their performance. At the strategic level the functional dependencies between task steps (pre-conditions) will be very important, because they are taken into account in planning.

There are two important aspects of the modes of control as part of the COCOM. One has to do with the conditions under which a person changes from one mode to another (keeping in mind that they only are points in a continuum); the other has to do with the characteristic performance in a given mode - i.e., what determines how actions are chosen and carried out. The first aspect is important for attempts to design and implement the COCOM as a dynamic tool, i.e., to make it operational. The second aspect is important for providing an account of the reliability of performance vis-a-vis the characteristic modes of control.

The change from one control mode to another, the transition between control states, can be described in many different ways ranging from the extremely simple to the extremely complex. A relatively simple solution is based on the use of three parameters:

o   Number of failed actions: this indicates the longest sequence of failed actions the model can cope with. If more failed actions occur, control changes to a lower level, e.g., from tactical to opportunistic.

o   Number of successful actions: this parameter controls the movement in the opposite direction. If a sequence of successful actions occur that is longer than this number, control changes to a higher level.

o   Number of goals accepted: this indicates the maximum number of goals the model can cope with on a given level of control. If more goals occur, for instance due to alarms or communication, the model will change to a lower level of control.

These rules must be combined with a pre-defined set of values for the parameters. To this must be added a description of characteristic performance on each level. One proposal is summarised as shown in Table 1.

	Strategic	Tactical	Opportunistic	Scrambled
Subjectively available time	Adequate	Adequate	Short or just adequate	Inadequate
Determination of outcome (cognition)	Elaborate, considering history + predictions	Normal, looking at main effects.	Concrete, limited by immediate effects.	Rudimentary
Choice of action	Prediction based	Plan based	Association based (dominant effects)	Random
Number of goals	Several	Several (limited)	One; or two competing goals	None (need dominated)

Table 1: Performance Characteristics for the Four Control Modes.

The performance differs with respect to how far ahead planning is made, how detailed the feedback is evaluated, the number of goals that are considered at the same time, etc. It is quite straightforward to apply these characteristics to the details of a specific scenario, and thereby provide a qualitative prediction of how adequate the operator's performance can be expected to be. It is a little more difficult to implement these characteristics as a running operator model, although by no means beyond what can be done with a reasonable effort. Such an effort is presently being made in the SRG project; the example being used is taken from aviation. The initial results are reported in e.g. [15].

4.3   Modes of Control and Reliability of Cognition

In COCOM the reliability of cognition is seen in relation to the four modes of control: scrambled, opportunistic, tactical, and strategic. It is assumed that the reliability of cognition is low for both the scrambled and opportunistic modes of control. In the scrambled mode of control the person is not considering the relevant system goals and the appropriate or prescribed procedures; actual performance will diverge from required performance and therefore be unreliable. In the opportunistic mode of control the person to some extent considers relevant goals and relevant plans, but the choice of action will be not always be the prescribed one. Cognition is driven by data (input information, events) rather than by intentions and cognitive goals; the match between desired and actual performance is therefore incomplete, and the overall result is reduced reliability.

In the tactical and strategic modes of control the reliability of cognition is assumed to be relatively high. In both modes there is little or no time pressure, and there is an adequate comprehension of the situation and the tasks. In neither case is the person likely to make major mistakes or miss events or decisions. In the tactical mode of control the person may return to a type of performance which is driven by rules and procedures, hence performed without constant and complete attention. It is consequently possible that the performance will not be completely adapted to the circumstances, hence that the outcome will not completely match the requirements. The reliability of cognition may be reduced because insufficient attention is paid to the ongoing variations in the environment.

In the strategic mode of control we must assume that the reliability of cognition is high - but not perfect. The limiting factor is the capacity of human cognition. In the strategic mode of control the person will try to consider as many relevant aspects of the situation as possible and choose an action that is a good compromise between needs and possibilities. But if the person tries to take too many details into consideration the ambition is self defeating: the result will be of a lower quality in terms of e.g. misunderstandings, incomplete reasoning, mistakes, reliance on heuristics, etc.

5   Conclusions

Analysis requires simplification. Some simplification is necessary, since one otherwise would be dealing with a replica of the real system. But it is important to think of a type of simplification which simplifies the analyst's work without at the same time simplifying the system description - at least not to the level usually done in static analyses. It is also important to be clear about why the simplification is made as well as which types of simplification are reasonable and which are not.

The guideline for this may be found in the Law of Requisite Variety. For the purpose of human reliability assessment it is necessary to capture how system development depends on the current conditions - not the ones which existed when the process started, but the ones that are actually present. One promising way of achieving this is to use simulations. The simulation approach has been pioneered by DYLAM and extended by making the models more sophisticated. The inclusion of a model of the human operator has been demonstrated by the CES [16], and brought a step further - though in a different way - by the SRG.

In the case of human performance the problem is that a reliability analysis of human cognition cannot be done by the same means as a reliability analysis of a technological system. The means for describing or modelling humans in detail sufficient to be used in a static analysis are simply not available. Paradoxically, it is actually easier to do it dynamically, by simulation.

If the solution is to make the model of the human operator dynamic, the question is how precisely this can be done. In the case of procedural prototypes the only option is by running through it. But that essentially means repeating the same sequence (the prototype) again and again. If it was wrong or insufficient from the start, it will not gain anything by being animated. The solution proposed here is to take a step backwards, so to speak, and consider the essential aspect of actions: control and choice. Consequently, the modelling will be concerned with how actions are controlled, rather than how the running through a prototypical procedure is regulated. The COCOM was proposed as a particular example of that. This further made it clear that it may not be necessary to have precise models of the details of cognition (micro-cognition); it may suffice to be able to simulate representative patterns of behaviour and / or modes of control.

The relation between COCOM and the reliability of cognition is important in two different ways. Firstly, if it is possible to analyse human performance in terms of cognition as described by the COCOM, then it may also be possible to produce results which are more consistent than by conventional methods. Secondly, if we can describe or identify specific conditions where the reliability of cognition is high, then this can be used as a basis for designing better systems. The ultimate goal of an analysis of human reliability is to identify the weaknesses in the system; it is not to provide numerical estimates of their size but to use them as a basis for improved design [17]. The simulation-based analysis of the reliability of cognition contributes to that.

Acknowledgments

The SRG project is part of the CEC DG XII programme on Science and Technology for Environmental Protection (STEP). The project is carried out by a consortium consisting of CRI A/S (prime contractor), Aerospatial Protection Systèmes (Paris, France), and the Institute for Systems Engineering and Informatics (Varese, Italy). The contributions of the members of the consortium and the support of the CEC are gratefully acknowledged.

References

1. Rosness, R., Hollnagel, E., Sten, T. & Taylor, J. R. Human reliability assessment methodology for the European Space Agency (STF75 F92020). Trondheim, Norway: SINTEF, 1992.

2. Park, K. S. Human reliability. Amsterdam: Elsevier, 1987.

3. Swain, A. D. & Guttmann, H. E. Handbook of human reliability analysis with emphasis on nuclear power plant applications. NUREG/CR-1278. Sandia National Laboratories, NM., 1983.

4. De Mey, M. The cognitive paradigm. Dordrecht: Reidel, 1982.

5. Hollnagel, E. & Woods, D. D. Cognitive systems engineering: New wine in new bottles. International Journal of Man-Machine Studies, 18, 1983, 583-600.

6. Woods, D. D. & Roth, E. M. Cognitive systems engineering. In M. Helander (Ed.), Handbook of human-computer interaction. Amsterdam: North-Holland, 1990.

7. Popper, K. The logic of scientific discovery. London: Hutchinson & Co, 1972.

8. Hollnagel, E. Reliability of cognition: Foundations of human reliability analysis. Loncon: Academic Press, 1993.

9. Hollnagel, E., Cacciabue P. C. & Rouhet J.-C. The use of an integrated system simulation for risk analysis and reliability assessment. 7th International Symposium on Loss Prevention, 4-8 May, 1992, Taormina, Italy.

10. Mancini, G. Modelling humans and machines. In E. Hollnagel, G. Mancini & D. D. Woods (Eds.), Intelligent decision support in process environments. Berlin: Springer Verlag, 1986.

11. Amendola, A. Accident sequence dynamic simulation versus event trees. In G. E. Apostolakis, P. Kafka & G. Mancini (Eds.), Accident sequence modeling. London: Elsevier Applied Science, 1988.

12. Bainbridge, L. Mental models in cognitive skill: The example of industrial process operation. In A. Rutherford & Y. Rogers (Eds.), Models in the mind. London: Academic Press, 1991.

13. Reason, J. T. Human error. Cambridge, U.K.: Cambridge University Press, 1990.

14. Rasmussen, J. Information processing and human-machine interaction: An approach to cognitive engineering. New York: North-Holland, 1986.

15. Cacciabue, P. C., Cojazzi, G., Hollnagel, E. & Mancini, S. Analysis and modelling of pilot-airplane interaction by an integrated simulation approach. IFAC, 1992.

16. Woods, D. D., Roth, E. M. & Pople, H. Jr. Modeling human intention formation for human reliability assessment. In G. E. Apostolakis, P. Kafka, & G. Mancini (Eds.), Accident sequence modelling: Human actions, system response, intelligent decision support. London: Elsevier Applied Science, 1988.

17. Moray, N. Dougherty's dilemma and the one-sidedness of human reliability analysis (HRA). Reliability Engineering and System Safety, 29, 1990, 337-344.

Human Errors in Dynamic Process Systems

John Wreathall

John Wreathall & Company, Inc., Dublin, Ohio 43017, USA

Abstract. Human error has been identified as a leading contributor to accidents and other abnormal occurrences in many fields, including those of aviation, chemical process safety, and medicine as well as nuclear power operations. Yet we are only starting to analyze the different dimensions that result in this large contribution. In particular, the nature of human errors and their interaction with systems in dynamic processes has not previously been placed on the research agenda.

Keywords. Human error, human reliability, human performance modelling, safety, disasters.

1   Introduction

The history of accidents and their analysis is also the history of human contribution to accidents. Increasingly through the industrial revolutions in the USA and Europe, "man-made" accidents started to become more widespread and with greater consequences. Rolt, for example, has provided accounts of British rail disasters (many of which were human error-related) dating from the beginning of of the industry in 1830 [1]. However, more formal consideration of the human contribution to accidents was reflected in the growth of the study of ergonomics in aviation during the Second World War. As technology has become more complex, often in parallel with a greater potential for harm to the public, consideration of the potential problems in human performance has increased. One of the first major process plant accidents that identified human errors as a principal cause was the Flixborough explosion in England on June 1, 1974 [2], where a design modification was inadequately developed because of a rush to continue production. Similar situations have shown themselves since in the petrochemical and process industries [3]. The accident at Three Mile Island, Unit 2, on March 28, 1979, was perhaps the first in the nuclear power industry that led to a critical review of all immediate aspects of the human factors environment: procedures, training, and the human-machine interface [4]. (A review of the human-machine interface issues in nuclear power control rooms had, however, been published shortly before the TMI-2 accident [5].) These three human-factors aspects mirrored the influences (the "performance shaping factors") analyzed in the human reliability technique used almost exclusively at that time: the Technique for Human Error Rate Prediction (THERP) [6].

A series of accidents took place in the 1980's that pushed the frontiers of human performance issues back from the immediate human causes and influences to include more of the organizational framework. These included the release of methyl isocyanate from Bhopal (1984), the destruction of the Space Shuttle *Challenger* during launch (1986), the accident at Chernobyl (1986), the capsizing of the passenger ferry *Herald of Free Enterprise* off Zeebrugge (1987), the fire in the London Underground station at Kings Cross (1987), the explosion on the North Sea oil platform *Piper Alpha* in 1988, and the British Rail accident at Clapham Junction, London, also in 1988. In all of these cases, much of the discussion of human failures has been directed at organizational issues. In fact, for the most recent cases, the official investigations have dwelt on the failures of the organization rather than of the individuals involved. For example, in the case of Zeebrugge, the formal report states " ... a full investigation of the circumstances of the disaster leads inexorably to the conclusion that the underlying or cardinal faults lay higher up the company. The Board of Directors did not appreciate their responsibility for the safe management of their ships." [7]. Most recently, in its report on the explosion and subsequent fires at the Phillips 66 Company facility in Houston in 1989, the U.S. Occupational Safety and Health Administration states simply "The primary causes of the accident were failures of the management of safety systems at the Houston Chemical Complex"[8]. Risk analysis has, to date, not actually addressed these organization factors, though work is proceeding in this area [9].

However, it is becoming increasingly recognized that the human contribution to disasters is only partly specified by the human factors and organizational issues. One part missing is the context in which the actions take place. This is the area in which work is now starting.

2   Framework of Accident Causation

Figure 1 presents the framework of accident causation that forms the core of this discussion; it is heavily influenced by that given by Reason in [10]. The essence of this framework is that when humans perform unsafe acts that breach defenses (or occur in the absence of defenses), accidents can occur. These unsafe acts are those actions (or even lack of actions) by people that are the operators and maintainers of the equipment, that cause a potentially dangerous condition. In well-protected systems, defenses exist to prevent these unsafe acts from becoming what Perrow called incidents and accidents [2]. These defenses can include hardware (safety systems, interlocks) and administrative controls (procedures, rules of conduct).

The unsafe acts do not take place in a vacuum. Rather the characteristics of the ergonomic environment play an important role in shaping the forms and probabilities of the unsafe acts. These factors can be job- and task-specific. However general principals for the effects of these factors have been developed and presented in various sources (discussed below).

At the upstream end, playing an important role in providing both the ergonomic environment and the effectiveness of installed defenses, lie the organizational processes, such as allocation of resources, and the setting of work policies and practices. Here are the wellsprings of good or bad safety potential.

2.1 Unsafe Acts

As far as the plant hardware is concerned, unsafe acts can be failures to perform actions to maintain the defenses-*errors of omission*-such as failing to start emergency equipment, or actions that cause or exacerbate the abnormal event-*errors of commission*-such as initiating a sequence of events. These unsafe acts can be *active* (that is, their consequences are immediately revealed as with an initiating event) or *latent* (that is, their consequences lay dormant in the system until triggered by some event as with deficiencies in maintenance). From the human point of view, there are a variety of unsafe acts: slips/lapses, mistakes, and circumventions. *Slips and lapses* are unsafe acts where what was performed was not what was intended, as with misselecting a control or skipping a step in the procedure. Mistakes are failures where the intentions are erroneous, but are purposefully executed. For example, a misdiagnosed failure in a component will result in a repair that is irrelevant to the failure mode; that is a mistake. The third category, circumventions, are deliberate but non-malicious breaches of safety rules. These are often done for "good" reasons like performing a task quickly or to overcome some organizational barrier. These categories are summarized in Figure 2, which provides examples of different kinds of each; they are discussed further in [10].

In terms of human reliability analyses performed in U.S. nuclear power settings, the emphasis has been on slips and lapses resulting in errors of omission; this is the principal emphasis of the THERP technology and its derivatives. Some attempts to evaluate mistakes as errors of omission in the context of misdiagnosis during post-accident recovery actions have been made using such techniques as the Operator Action Tree (OAT) method [11]. However, these can only be considered as interim (and were so when developed by their authors!).

2.2 Error-Producing Conditions

Referring to Figure 1, the improvement in understanding in the middle part of the 1980's was in the area of the error-producing conditions-the so-called performance shaping factors of THERP. In that method, the principal factors include format and content of procedures, adherence to administrative rules, layout and style of displays, item labeling and tagging, stress, and the effectiveness of checking. Alternatively Williams developed an extensive database for quantifying the effects of various error-producing conditions based on a review of published data; some of these conditions are summarized in Table 1 [12].

An alternative structure of the error-producing conditions has been presented by Wreathall, et al [13]. This structure, termed the onion model, has been developed to portray a more diverse set of influences on job performers that includes traditional ergonomic issues such as human-factors engineering, training, and environment. It also includes psychological and sociological factors, such as team structure, professionalism, policy consistency, and the rewards/punishment structure of the work unit. No formal attempt has been made to quantify the strengths of influences, since

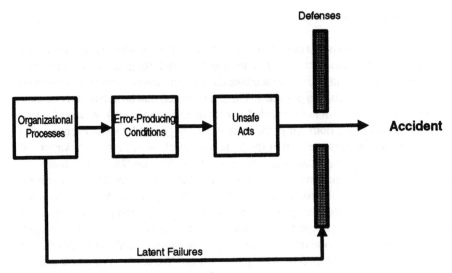

Fig. 1. Framework of accident causation

it is recognized that the factors are not well understood and their influences are contingent on many factors both internal and external to the work unit.

One factor of note is that the ergonomic environment can be simultaneously good and bad. For example, in a review of human performance problems during low power and shutdown operations for U.S.NRC, a simple count was made of the number of times various different elements of the ergonomic environment contributed to unsafe acts and to recovery actions. An example of one such case is shown in Figure 2. Of interest is the fact that the same factor, such as procedures contributes to both good and bad. This suggests that the development of human reliability analysis models cannot be based on such factors, or at least this level of specification. This finding is discussed further below.

2.3   Organizational Processes

It is recognized that only a limited amount of work has been performed on evaluating how organizational processes influence safety. Noteworthy as early contributors are the work by Westrum on identifying the characteristics of safe organizations [14], and that by the University of California, Berkeley, group on high reliability organizations [15]. Westrum describes the characteristics of organization behavior as generative, calculative, or pathogenic. Generative organizations are those that accomplish high levels of success by apparently unconventional of exceptionally high expectations. Hazards are identified and removed by "lower level" personnel empowered to seek out and eliminate problems quickly. Calculative organizations perform functions by the book in conventional ways, often meeting regulatory requirements but rarely exceeding them. Cost-benefit is typically a byword for decisionmaking. The final group, pathogenic organizations, typically consider safety regulations as a barrier to

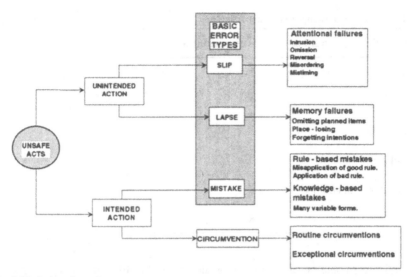

Fig. 2. Varieties of unsafe acts

production. Messengers bearing safety warnings are typically shunted to one side, ignored, or fired.

The work on high reliability organizations describes examples of how organizations fitting Westrum's generative category practically accomplish their performance. These include case studies from aircraft carrier flight-deck operations, air traffic controllers, and electrical power dispatchers. Issues identified in this work include: empowerment of junior staff to stop production, with a reward-based support system for those who do; a transformation process of responsibilities such that the most experienced personnel take control when things go wrong; and that the safety "boundary" is being monitored continuously by several people ready to declare an incident.

Different views of organizational processes have been presented by Reason [16] and Wreathall [17] in terms of organizational "core processes"; that is, those activities that must be accomplished within the organization to manage safety. Reason has described the four basic processes of a technical system as comprising Design-Build-Operate-Maintain (D-B-O-M), with management and communication acting between them. (In today's environment, decommissioning must be added-D-B-O-M-D.) While these apply at the working level, there are two additional processes that contain the technical processes. These are Goal Statement and Organization. Beyond the organization is an additional process: regulation. Using this concept, Reason then develops a set of eleven Organizational Failure Types (OFTs) that would comprise evidence of inadequacies in these processes. These include: incompatible goals, inappropriate organizational structure, inadequate communications, poor planning and scheduling, poor procedures, poor training, and so on. Once identified and interpreted in an operational context,

184

Table 1. Error producing conditions (taken from [12])

Error-Producing Condition	Maximum Influence
Unfamiliarity	x 17
Time shortage	x 11
Low S/N ratio	x 10
Interlocking	x 9
Functional incompatibility	x 8
Model mismatch	x 8

these indicators could be used to monitor the safety "health" of the organization processes.

Wreathall [18] proposed a different view of core processes as being the fundamental activities hat must be performed similarly at all levels of the organization, the regulatory environment, and beyond. These are: competence, commitment, and awareness. At the working level, individuals internalize all three processes and manage the risks associated with their own individual activities. At higher levels of the organization, these processes become separated, with line management providing commitment (of resources), safety professionals providing awareness, and the technical staff providing competence (through training, etc.). The effectiveness of these functions at different levels can be monitored and managed in different ways using performance indicators as a means for evaluating performance. Issues in this framework include identifying who is responsible for the different processes at different levels and what are appropriate performance indicators .

Work is proceeding in this area of organizational processes and their impact on safety; see, for example, [9], [19], and [20] for recent developments.

3   The Importance of Context

The observation earlier, that the same factors in the ergonomic environment are both good and bad is an indication that the potential influence of human performance and human factors cannot be considered in isolation. As Reason and others have observed, bad things happen to "good" plants and "good" things happen to bad plants; safety has a large element of chance in it. Also, simply exhorting plants to improve procedures, training, and human-factors designs will not remove the opportunities for disasters, though it may reduce the chances by removing some, though not all, of the latent failures. Stated differently, there never will be perfect procedures, or perfect training, or any other perfect influence. Certainly, it is important that continuous improvements in these areas continue, using such methods as *kaizen* as described in [21].

However, the second part of the issue is the conditions under which such latent failures lead to a disaster. For example, what are the conditions that lead errors in procedures and training to become so serious? Organizations and their personnel cannot maintain perfect vigilance all the time. If it is possible to specify the conditions under which unsafe acts may result in accidents, then for those periods perhaps vigilance can be raised and compensatory measures put in place before the unsafe acts occur rather than after. An analogy can be drawn with a ship that must sail into an area of icebergs. Icebergs are well recognized as a peril to shipping. Therefore a prudent captain may well ensure that a speed reduction be made, that open accessways be closed, and that backup radar is available before entering the area. This is done because the danger is so self evident, not that regulations require such defenses at all times. For operating facilities (nuclear, chemical ,or aviation), what is required is the equivalent description of the conditions where risks are higher-this is the issue of context. Disasters are the result of unsafe acts in the context of ... . The new challenge is to complete the description of context so that, as Harbour discusses [22], proactive measures can be taken when the need is recognized.

Early into their CES project related to the reliability of operators in nuclear power plant control rooms, Woods, et al, [23] recognized that the definition of context provided by probabilistic risk assessments (PRAs) was entirely inadequate for analyzing human errors. Detailed representations of the control room displays, specific knowledge of operators, and time-sequencing of events were found to be important factors that are not described in PRAs yet, as the CES project demonstrated, these are critical factors in determining the probabilities of successful operator actions and can force operators into incorrect judgements. More recent findings of this project are presented in [24]. This work indicates that the assessment of human error probabilities that addresses only the influence of factors like procedure formats and panel layouts statically neglect the critical issue of how many times operators are driven to err by the environment. For example, if the conditions that force operators to err occurs once in a thousand times, the baseline human error rate cannot be below 10^{-3} per occurrence. Yet to identify the conditions that can force such an error is not a trivial task. No methods exist today on how to search for such conditions.

It is recognized that this is a new challenge to the human error, human factors, and human reliability communities. Initial work in different areas is starting. One concept being considered by Woods and others [25] is the body of six rules by Norman on how to create a completely unusable design [26]. These were originally developed in relation to personal computer systems. These are:

"Make things invisible. widen the Gulf of Execution; give no hints to the operation expected. Establish a Gulf of Evaluation: give no feedback, no visible results of the action just taken Exploit the tyranny of the blank screen.

Be arbitrary. Computers make this easy. Use nonobvious command names or actions. Use arbitrary mappings between the intended action and what must actually be done.

Be inconsistent: change the rules. Let something be done one way in one mode and another way in another mode. This is especially effective where it is necessary to go back and forth between the two modes.

Make operations unintelligible. Use idiosyncratic language or abbreviations. Use uninformative error messages.

Be impolite. Treat erroneous actions by the user as breaches of contract. Snarl. Insult. Mumble unintelligible verbiage.

Make operations dangerous. Allow a single erroneous action to destroy invaluable work. Make it easy to do disastrous things. But put warnings in the manual; then, when people complain, you can ask, "But didn't you read the manual?"

While these rules do not directly relate to the design of human actions in operating large hazardous facilities, there are general principles that do clearly apply, such as those of invisibility (adequacy of relevant information), and arbitrariness and inconsistency (job design and procedures). For example, during low power and shutdown events, these principles clearly play a role in the frequencies of incidents.

This must be recognized as a start. Clearly, these rules do not provide a clear and sufficient set of guidelines to identify when the context of work is becoming dangerous. However, if such guidelines can be developed, then it should be possible to ensure adequate defenses are installed before an incident, not recovery or mitigation actions after the accident. For example, as modes of operation change in a power plant (as in moving to low-power operations), it should be possible to recognize that the plant behavior is becoming less visible where temporary instrumentation has much less salience than at full power, that the rules of thumb for operation no longer apply (inconsistency), and that lines of command are becoming less clear (arbitrariness may increase as contractors are supervised in ways different from the plant staff).

4 Human Reliability and Dynamic Systems

There are at least two additional characteristics of human performance that need to be considered in evaluating the reliability and safety of dynamic process systems. First, HRA techniques presently in use consider (in most cases) a static view of the world. Specifically that while equipment failures and the plant condition may evolve in time, those failures do not cause further aberrational states that may further cause problems, which in turn cause more failures. Such a propagation may occur because of human or equipment failures, particularly in the control functions. As described in other papers in these proceedings, one of the most characteristic causes of reliability and safety problems in dynamic systems is the interaction of process and control systems under faulted conditions.

Human interactions with processes involving flawed controllers can be considered in two separate yet related issues. First, with failed hardware controllers, fault-finding and diagnostic activities by operators can be challenging. Operators may try a variety of manual interactions with the plant to find where are the failure modes. For example, if a level is inexplicably falling, operators may try closing valves in various drain paths, tripping pumps, and so on, to search for possible causes. If the failures are in the controlling system, it may react to these changes of state and operate other components to "compensate" for the manual changes. In this way, the fault finding may be prevented by seemingly erratic behaviors of the failed controllers.

This imposition of manual control on process interactions implies that human-system interactions to allow fault finding and system state recovery can be accomplished. (A simple example of such an ability is the designed-in capability of airplane pilots to "dump" the autopilot in the event of a malfunction and use entirely manual control.) This can be accomplished where manual operation can replace the automatic controller. However, in modern military airplanes, such as the Stealth airplanes, loss of automatic controllers means loss of the airplane. Presumably, the evolution of control systems in process plants will reach a similar threshold. Perhaps they already have.

The second concern in assessing the human reliability in dynamic process settings is identifying the potential consequences of human errors by operators in controlling the plant. If the plant is controlled dynamically by human rather than automatic controllers, then there are a wide range of possible failure states that could exist. Current HRA methods consider errors like omitting a step in a procedure or misdiagnosing a major accident (to take two examples). However, no methods exist for assessing the probability of controlling with "too wide a deadband" or with too long a "time constant". There are effects that may make people's skills in process control degrade, such as workload and fatigue, so that performance by one individual may well vary through time. These will be valid concerns for the assessment of human reliability in dynamic settings, for which no work has been started.

In addition to these specific HRA issues, safety is often accomplished by operation of separate safety equipment whose operation is independent of process control devices. This is commonly so in high-hazard industries like nuclear power and chemical manufacturing. However, production reliability and economic losses (equipment damage, etc.) can still be adversely affected. However, it is not clear that such a separation can be maintained indefinitely. As processes increase in complexity and scale, it may be that safety becomes increasingly reliant on operation of the controllers themselves, either because safety devices become too slow compared with the speed of reactions, or the complexity of the processes prevents simple safety devices from creating a stable state. The potential for design deficiencies that may act as common-mode failures must be considered

References

1. Rolt, L.T.C.: Red for Danger. London: Pan Books, 1978.
2. Perrow, C.: Normal Accidents: Living with High-Risk Technologies. New York: Basic Books, 1984.
3. Kletz, T. A.: What Went Wrong?: Case Histories of Process Plant Disasters. Houston: Gulf Publishing Co., 1985.
4. U.S. Nuclear Regulatory Commission, NRC Action Plan Developed as a Result of the TMI-2 Accident (NUREG-0660). Washington, D.C., May 1980.
5. Seminara, J.L., Gonzalez, W., and Parsons, S.: Human Factors Review of Nuclear Power Plant Control Room Design (EPRI NP-309). Palo Alto, Ca: Electric Power Research Institute, March 1977.
6. Swain, A.D., and Guttmann, H.E.: Handbook of Human Reliability Analysis with Emphasis on Nuclear Power Plant Applications (NUREG/CR-1278). Albuquerque, N.M.: Sandia National Laboratories, 1983.
7. Department of Transport (UK), mv HERALD OF FREE ENTERPRISE, Report of Court No. 8074, Formal Investigation. London: HMSO, July 1987, p 14.
8. U.S. Department of Labor, The Phillips 66 Company Houston Chemical Complex Explosion and Fire. Washington, D.C., April 1990.
9. See, for example, Apostolakis, G.: On the Inclusion of Organizational Factors into Probabilistic Safety Assessments of Nuclear Power Plants. IEEE Conference on Human Factors and Nuclear Safety, Monterey, CA, June 1992 .
10. Reason, J.: Human Error. New York: Cambridge University Press, 1990.
11. Hall, R.E., Fragola, J.R., and Wreathall, J.:Post Event Human Decision Errors: Operator Action Tree/Time Reliability Correlation (NUREG/CR-3010). Upton, N.Y.: Brookhaven National Laboratory, November 1982.
12. Williams, J.C.: "A Data-based Method for Assessing and Reducing Human Error to Improve Operational Performance," in 1988 IEEE Fourth Conference on Human Factors and Power Plants, Hagen, E.W. (Ed.). New York: Institute of Electrical and Electronics Engineers, 1988.
13. Wreathall, J., Schurman, D.L., and Anderson, N.A.: "An Observation on Human Performance and Safety: The Onion Model of Human Performance Influence Factors." in Proceedings of the International Conference on Probabilistic Safety Assessment and Management (PSAM), Apostolakis, G. (Ed.). New York: Elsevier Science Publishing Co. Inc., 1991.
14. Westrum, R.: Organizational and Interorganizational Thought. World Bank Workshop on Safety Control and Risk Management, Washington, D.C., October 1988.
15. See, for example, Roberts, K.H.: "Some Characteristics of High Reliability Organizations," Organization Science, vol. 1 no. 2, pp160-177, 1990. Also Rochlin, G., La Porte, T.R., and Roberts, K.H.: "The Self-designing High-Reliability Organization: Aircraft Carrier Flight Operations at Sea," Naval War College Review, vol. 40 no.4, pp76-90, 1987.
16. Reason, J.: "Disasters and Human Failures," in Psychological Aspects of Disasters, A. Taylor, D. Lane, and H. Muir (Eds.). Leicester (UK): British Psychological Society, 1991.
17. Wreathall, J.: A Hierarchy of Risk Control Measures, with Some Consideration of "Unorganizational Accidents". Second World Bank Workshop on Risk Management & Safety Control, Karlsdad, Sweden, November 1989.
18. Wreathall, J.: Indicators in Search of Safety, 1991 New Technolgy & Work (NeTWork) Meeting. Bad Homburg (Germany): Werner Reimer Stiftung, May 1991.

19. Haber, S.B.: The Nuclear Organization and Management Analysis Concept Methodology: Four Years Later. IEEE Conference on Human Factors and Nuclear Safety, Monterey, CA, June 1992 .

20. Jacobs, R.: Assessment of Organizational Factors in Nuclear Power Plants. IEEE Conference on Human Factors and Nuclear Safety, Monterey, CA, June 1992 .

21. Hallbert, B.P., and Harbour, J.L.: Developing Static Information Displays: A Case Study on Improving Worker Safety. IEEE Conference on Human Factors and Nuclear Safety, Monterey, CA, June 1992 .

22. Harbour, J.L.: Integrating HRA into Decision Support Systems: A New Frontier? IEEE Conference on Human Factors and Nuclear Safety, Monterey, CA, June 1992 .

23. Woods, D.D., Roth, E.M., and Hanes, L.F.: Models of Cognitive Behavior in Nuclear Power Plant Personnel (NUREG/CR-4532). Pittsburgh, Pa: Westinghouse Research and Development Center, July 1986.

24. Roth, E.: Lessons Learned on Cognitive Environment Simulator Program-Severe Accident. IEEE Conference on Human Factors and Nuclear Safety, Monterey, CA, June 1992 .

25. Cook, R.I., Woods, D.D., and Howie, M.B.: "Unintentional Delivery of Vasoactive Drugs with an Electromechanical Infusion Device," in Journal of Cardiothoracic and Vascular Anesthesia, vol. 6, no. 2, April 1992, pp 238-244.

26. Norman, D.A.: The Design of Everyday Things. New York: Doubleday/Currency, 1990, pp 178-179.

Part 4

Risk Assessment for Realistic Decision Making

Risk Assessment for Realistic Decision Making: An Overview

P.Carlo Cacciabue

Commission of the European Communities, Joint Research Centre, Institute for Systems Engineering and Informatics, 21020 Ispra (Va), Italy

Abstract. The role of dynamic approaches to risk assessment for realistic decision making in dynamic systems is discussed with regard to conventional approaches.

Keywords. Dynamic methodologies, risk assessment, human factors, fault tree/event tree approach, system simulation

1 Introduction

During incidental sequences, the effect of the plant dynamic evolution on the failures of components can lead to non-intuitive configuration of the system and consequently invalidate a safety analysis of the plant, unless a representation of the physical behaviour of the plant transient is called in support of such analysis. This issue becomes crucial when the consideration for the interaction of the operator and the plant control and protection system is entered in the overall scenario of the safety study and, in particular, in all cases of development of Probabilistic Safety Assessment (PSA). These considerations have always been recognized and the use of simulation codes (US NRC, 1981) and human decision making analysis (Swain and Guttman, 1983) have been strongly recommended in order:

1. to acquire deep knowledge of the control process, and,

2   for developing advanced methods of risk assessment.

However, in practice, for realistic decision making when using PSA techniques, the analysis reduces to the definition of success and failures of different systems expected to intervene during the transient. This means that, usually, there is an oversimplification of the actual evolution of the physical and mental processes sustaining the behaviour of plant components and operator decisions.

These problems, which represent the driving issue behind the whole exercise of the NATO Advanced Research Workshop on the Reliability and Safety Assessment of Dynamic Process Systems, are common to all the practical domains in which the use of PSA represent a consolidated approach for the safety analysis. In particular, when realistic decisions have to be taken at design level as well as verification stage, the collection of a number of practical viewpoints from the industries, as well as from the safety authorities, are of great interest for the future research and development in this domain. The industries mostly exploiting the PSA methods are the nuclear power and aerospace even if also in the chemical and petro-chemical domain the technique is becoming more commonly adopted (Apostolakis, Ed., 1991).

2 Risk assessment for Realistic Decision Making

Even if the theoretical aspects of the application of the PSA methods is outside the scope of the present analysis, an overview of the state of the art of dynamic approaches and their ability to tackle and solve real problems in PSA and risk analysis is necessary in order to put the problem into perspective (Hirschberg, 1992). Indeed, although the problem has been and is still being recognized as important and actual, the possibility to extent the currently developed methods to the whole PSA analysis is still questionable for a number of reasons such as the complexity of a real plant configuration, the difficulty to describe in detailed terms the ongoing phenomena and the amount of data needed.

From a practical perspective, an important exercise is to start from the application of "classical" fault trees/event tree techniques and try to derive the role of alternative methods, when a real plant system is the object of the study. This approach is relevant because, usually, the "dynamic" techniques are applied to benchmark or sample case exercises. In the nuclear power production domain, for example, the role and opportunities of application of methods, alternative to fault trees/event trees during the development of a PSA, has been extensively discussed by Bley (1992), while Kafka (1992) has analysed the practical application of "classical" approaches, like large fault trees method, for the study of dynamic system behavior within PSA applications. Similarly, in the aerospace domain the methods of digraphs and of fault trees have been demonstrated to represent a viable option for studies in aerospace safety, in particular for the NASA research (Patterson-Hine, 1992)

The analysis of the role and position of the regulatory body, which aims at bridging the gap between the researchers and practitioners of PSA methods and the layman, shows that the use and application of "classical" methods is becoming now a common practice (Potter, 1992), while the application of dynamic techniques still needs to be more detailed and better focused before representing a valid alternative option to the fault tree/event tree approach.

3 Conclusion

The review of methods and techniques adopted in real applications and realistic decision making when using the PSA approach seems to consolidate the role and importance of "classical" techniques, even if their application presents limitations and drawbacks. A general recognition exists concerning the needs for dynamic methods, and the amount of information derived from their application is accepted to be much more accurate than the one obtained with the fault tree/event tree approach. Nonetheless, the existing approaches do not seem to have the necessary impact on the domain of actual plant analysis, either because of their complexity or because they need to be supported by enormous amount of data and/or simulation programs. Moreover, the dynamic methods have been demonstrated and applied almost always to study cases of limited complexity, while it is still to be demonstrated their full applicability to a real plant situation as well as the exploitation of the information derived from their application for PSA and safety analysis.

References

Apostolakis, G. (Ed.) (1992). Proceedings of International Conference on Probabilistic Safety Assessment and Management. Beverly Hills, California, 4-7 February 1991. Elsevier, New York, NY.

Bley, D. (1992). Advanced Modelling of Dynamic Process Systems: Where are Targets of Opportunity in PSA? These Proceedings.

Hirschberg, S. (1992). Time Dependencies in Probabilistic Safety Assessment. These Proceedings.

Kafka, P. (1992). Approximations of the Dynamic System Behavior within the Process of PSA. These Proceedings.

Patterson-Hine, A. (1992). Object Representation of Common Reliability Models: New Solution Techniques and Extensions. These Proceedings.

Potter, S. (1992). The Approach of a Regulatory Body to Assessment of Risks from Major Hazard Chemical Plants. These Proceedings.

Swain, A. D. and H.E. Guttman (1983). Handbook on Human Reliability Analysis with Emphasis on Nuclear Power Plant Application. Draft Report.

NUREG/CR-1278. SAND 80-0200 RX, AN. Final Report. Sandia National Laboratories, Albuquerque, New Mexico.

Time Dependencies in Probabilistic Safety Assessment

Stefan Hirschberg[1] and Michael Knochenhauer[2]

[1]Paul Scherrer Institute, CH-5232 Villigen PSI, Switzerland
[2]Logistica Consulting AB, Västerås, Sweden

Abstract. The paper uses as a starting point a classification of different types of time dependencies encountered in Probabilistic Safety Assessment (PSA). This includes: time-dependent failure rates (ageing, learning), time-dependent unavailabilities (test interval and test arrangement dependencies, latent failures not revealed in tests), time dependencies of accident sequences (time-dependent success criteria, timing of safety system operation, timing of operator actions, time-dependent operator error and recovery probabilities, time-dependent physical phenomena), and increase of statistical evidence with increasing operational experience. The currently proposed dynamic methodologies are intended to improve modeling of time dependencies associated primarily with accident sequences; some of the other above mentioned dependencies are also implicitly present in sequences of interest and their treatment may be very important. This is illustrated by some case studies originating from projects performed in Nordic countries. An example of an important sequence from the Swedish PSAs is described. The sequence was identified as dominant for the latest generation of operating Swedish BWRs; it involves several of the different types of time dependencies. The modeling approaches used are based on simplifications and more detailed/realistic representation offered by the dynamic methodologies, would be welcome. Finally, comments are given with regard to the problem-solving potential offered by the new approaches. Some questions concerning the practical aspects of the dynamic methodologies are raised.

Keywords. Time dependencies, classification, examples, assumptions, impacts, decision-making, dynamic methodologies, need, potential.

1 Introduction

Numerous time-dependent phenomena can have impact on the results of Probabilistic Safety Assessments (PSAs) and on decisions based on the outcome of such studies. The modeling of time dependencies in current PSAs is predominantly based on static approaches, averages and/or relatively crude simplifications of the real world to approximate the complex system behaviour.

Thus, in the conventional event trees the systems functions and operator actions are supposed to roughly reflect the expected sequence of events. This representation does not allow for detailed representation of the actual interactions between the evolution of the physical process, the system hardware configuration and operator behaviour. In addition, the systems models, i.e. fault trees, are static models of the systems' unavailability. The reason behind the current research on dynamic methods is the potential of such approaches to simulate the actual plant response and consequently overcome some of the intrinsic limitations of traditional methods. The dynamic approaches can be roughly divided into those based on analytical techniques (Markov models; e.g. Papazoglou 1992) and on simulation (dynamic event trees and Monte Carlo; e.g. Acosta and Siu 1991, respectively Marseguerra and Zio 1992).

While it is clear that in principle all systems are dynamic and use of static methods is an attempt to approximate their true behaviour, there are some differences of opinion with regard to how good such approximations are. Generalizations are difficult since the effects may be significantly different from case to case. Traditionally, the practitioners' point of view is that in most cases the approximations are good enough. The usual PSA praxis in special cases where conventional approaches to time dependencies are obviously inadequate, is to give them a separate and pragmatic treatment. However, such representation of time-dependent effects is rather crude when compared with the conceptually much more "exact" but complex and time/resource consuming dynamic approaches.

When evaluating the potential benefits of dynamic methods it is natural to ask what is the relative importance of those PSA limitations that can be resolved/minimized by introduction of dynamic approaches. Some of the main PSA limitations have been recently discussed by one of the authors from qualitative and quantitative point of view, respectively (Hirschberg to be published, Hirschberg 1990). In the present paper the focus is on one specific limitation of PSAs, i.e. treatment of time dependencies. The purpose is to focus the attention on the fact that the dynamic methods may be capable to improve the treatment of *specific types of time-dependent phenomena* in PSAs. At the same time there are other types of time dependencies whose treatment will not be resolved by introduction of dynamic approaches. The difficulties in treatment of such time dependencies are directly related to the limitations of available data and the simplifying assumptions made in the process leading to generation of data. There is no intention to provide in the present paper a complete review of this issue; instead some characteristic examples are described. The same applies to some cases of typical decision-making situations involving time dependencies. The fact that the problems described are serious and will hardly be solved by the dynamic approaches should not be taken as a criticism of those. The intention is to provide a balanced perspective on the potential capabilities of dynamic approaches. An important accident sequence with complex time dependencies involved is described. It could be considered as a possible candidate for testing the capacity of dynamic approaches.

2 Time Dependency Categories

A simplified classification of time dependencies encountered in PSAs follows below. For a more detailed discussion we refer to (Hirschberg ed. 1990).

1. Time-dependent failure rates

   — ageing
   — learning

   This category affects component failure probabilities and initiating event frequencies associated with failures of major passive components (piping, reactor pressure vessel). Standard PSAs disregard the effects of ageing and learning. Lack of empirical data and practical models are the reasons behind this assumption, although the phenomena (particularly ageing) are sometimes considered in sensitivity studies. New developments are underway; for example, methodology for handling age-related phenomena has been developed (Vesely and Hassan 1991).

2. Time-dependent unavailabilities

   — test interval dependencies
   — test arrangement dependencies
   — latent failures not revealed in tests

   The unavailability of a standby system is highly dependent on the test interval, the test arrangements and test efficiency. These types of time dependencies were studied in detail in a recent Nordic research project (Laakso et al. 1990) and examples of some findings will be given in the next section of the present paper. In standard PSAs the unavailability of a standby system is usually evaluated as mean unavailability, which is test interval dependent. In some cases the (calendar) time-dependent unavailability is required. Such cases may include situations with long inspection intervals and/or low test efficiency. Also timing and actual arrangements of the tests may be very important for example in the context of revealability of common cause failures, with corresponding impacts on system unavailability. The application of complex time-dependent models in PSAs is partially tempered by the limitations of standard fault tree codes, but mainly by lack of detailed data.

3. Time dependencies of accident sequences

   — time-dependent success criteria
   — timing of safety system operation
   — timing of operator actions
   — time dependency of operator error probabilities

— time-dependent recovery probabilities
— time-dependent physical phenomena

This category of time dependencies is central for accident propagation modeling. It is in the context of the treatment of this particular and very important type of time dependency that the novel dynamic approaches could bring significant improvements. The potential gains are foreseen both in the context of better understanding of accident propagation (with possible identification of specific scenarios not clearly identified today due to more limited resolution of current much less detailed approaches), and with respect to higher precision of numerical results. It should be noted, however, that the previously mentioned types of time dependencies are implicitly present in the modeling of accident sequences and result in corresponding limitations of the approaches used (independently whether they are dynamic or not). The difficulties associated with the treatment of the time dependencies belonging to this group are well known and described in standard literature. It is worth noting that some of the aspects (for example relaxation in success criteria with time and impact of recoveries) become even more important in PSAs covering shutdown and low power operation.

4. Increase of statistical evidence

Two issues are of interest here. First, the increased evidence/knowledge can reveal incompabilities between models and actual experience. Second, the new operating experience needs to be implemented by updating the PSA in the spirit of the "Living" PSA concept.

3 Some Typical Approximations and Impacts

The process leading to generation of component failure data involves a number of simplifications and assumptions, the most important ones being (Knochenhauer 1989):

— periodical testing is assumed to cover all relevant failure modes of the component, i.e. all latent failures are assumed to be discovered;

— periodical testing is assumed to cover all relevant operating modes of the component;

— failures discovered in periodical testing are assumed to be efficiently eliminated;

— no failures are assumed to be introduced through periodical testing;

— the failure reporting is assumed to be adequate, i.e. all failures are reported, all reports are correct and all "test" events are recorded;

— the failure reporting is usually symptom-oriented whereas the data analyst would be better helped by cause oriented reporting;

— by necessity many of the failure data used are based on generic information which may introduce systematic errors when the data are used for specific applications;

— to obtain reasonable population sizes, components are usually grouped, which will sometimes introduce additional uncertainties.

The problems outlined above are summarized in Fig. 3.1.

A FRANTIC-based sensitivity analysis performed for the high pressure injection system of a Swedish BWR (Knochenhauer 1988b), illustrates the impact of assumptions concerning test characteristics on the estimated system unavailabilty . The following parameters were varied in this study:

— test scheme (sequential/staggered)
— test inefficiency (P_e)
— test introduced failures (P_i)
— test override unavailability (P_o)
— test duration (τ)

Fig. 3.2 shows some of the results of the study in terms of relative changes in the mean system unavailability as compared to the reference case (PSA for the plant analyzed, characterized by staggered testing and assumptions of perfect tests). Obviously, the results are very sensitive to the value of the test inefficiency parameter. This is due to the fact that the component failure probability is assumed to be test interval-dependent.

The actual relationship between the component unavailability and the test interval was investigated in a detailed study for motor-operated valves (MOVs) at nine BWRs of ABB Atom design, operating in Sweden and Finland (Knochenhauer 1988a); the study covers 78 reactor years. The results are displayed in Fig. 3.3 and demonstrate the validity of the linear standby failure rate model. It is worth noting that corresponding evidence supporting this model has been demonstrated only for few component types. At the same time a number of PSA applications, particularly those supporting plant operation (for example optimization of surveillance test intervals), are directly dependent on the validity of such a model (at least that is the assumption generally used in current applications).

For a survey of related topics concerning coverage of testing, applicability of data originating from tests to real demand situations and quality of failure reporting we refer to (Knochenhauer 1989).

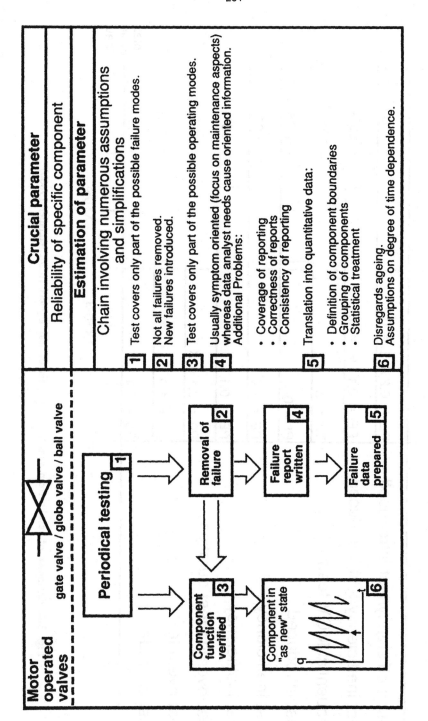

Fig. 3.1. Approximations involved in estimation of component reliability (Knochenhauer 1989)

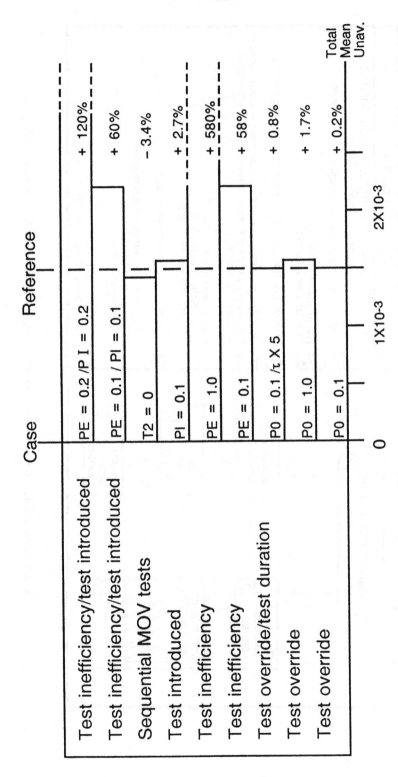

Fig. 3.2. Examples of results from sensitivity analysis of test aspects (Knochenhauer 1988b)

Plant level

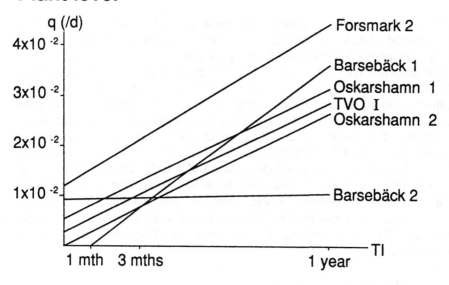

Total results

$$q\,(TI) = 1.4 \cdot 10^{-4} + 2.8 \cdot 10^{-6} \cdot TI$$

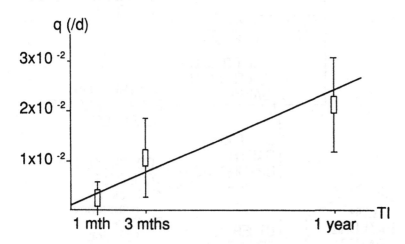

Fig. 3.3. MOV failure per demand probability as a function of test interval (Knochenhauer 1988a)

4 Time Dependencies and Decision-Making

In (Knochenhauer and Hirschberg 1991) probabilistically based decision support is addressed. Here some examples originating from Nordic research studies and directly involving time dependencies are presented.

The safety influence of failures in standby safety systems can be considered from the point of view of: (a) instantaneous risk frequency; (b) expected risk over failure situation; (c) contribution to the total risk over plant lifetime. All these measures need to be considered when defining criteria for repairs and action statements. The criteria approach is structured by a decision tree provided in Fig. 4.1 (Laakso et al. 1990). Detailed treatment of operational decision alternatives in failure situations (including multiple failures) of standby safety systems can be found in (Mankamo and Kosonen 1991). It has been demonstrated that in specific failure situations involving residual heat removal systems, the shutdown constitutes a higher risk than continued operation over repair times of normal duration. This has resulted in appropriate modifications of Technical Specifications.

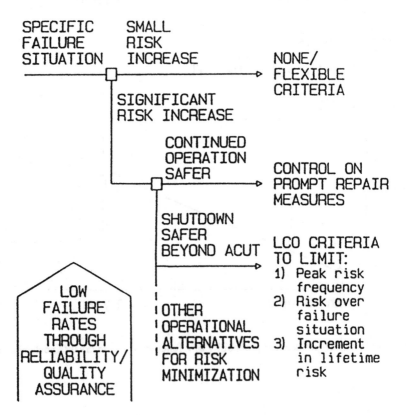

Fig. 4.1. Decision tree for criteria on Allowed Outage Time (AOT) during power operation in the case of critical failures detected in safety systems (Laakso et al. 1990)

Interesting decision situations occur when considering the acceptable extent of preventive maintenance activities during operation. In the latest generation of Nordic BWR plants with four train safety systems, preventive maintenance (PM) during operation is currently allowed in one subsystem at a time with a budget criterion. The calculated annual average risk increase is a few percent not taking into account the obvious benefits of PM, that are not easily quantified (Knochenhauer 1987). The temporary risk increase has been minimized by grouping PM into functionally linked subsystems, and by excluding simultaneous maintenance on subsystems in redundant safety systems. The small contribution illustrated in Fig. 4.2 (on the next page), can be explained by the high level of redundancy and the dominance of high multiplicity common cause failures (CCFs).

Another type of decision situations involving time dependencies is associated with the new developments in uses of PSA for risk monitoring/risk follow-up (Holmberg et al., 1992). Fig. 4.3 illustrates the basic frequency curves. The baseline risk f_0 is the theoretically lowest attainable frequency level (no components unavailable due to maintenance, all standby components just tested). The instantaneous risk f(t) varies around the nominal risk f_n, and $f_i(t)$ is the discretized instantaneous risk. Risk monitoring is supposed to reflect the current plant configuration. Risk follow-up calculates retrospectively the actual risk experienced during the operation of the plant. In principle risk monitoring should be performed on-line, which leads to requirements on short response times. In its most advanced form risk follow-up takes into account both failures and successes of past tests, which complicates the time-dependent model to be applied.

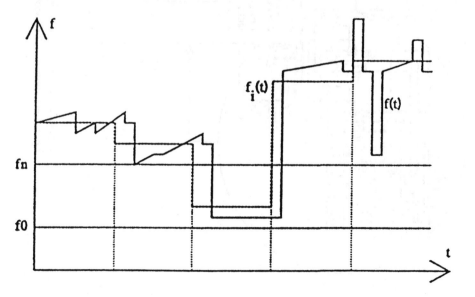

Fig. 4.3. The basic risk frequency curves (Holmberg et al. 1992)

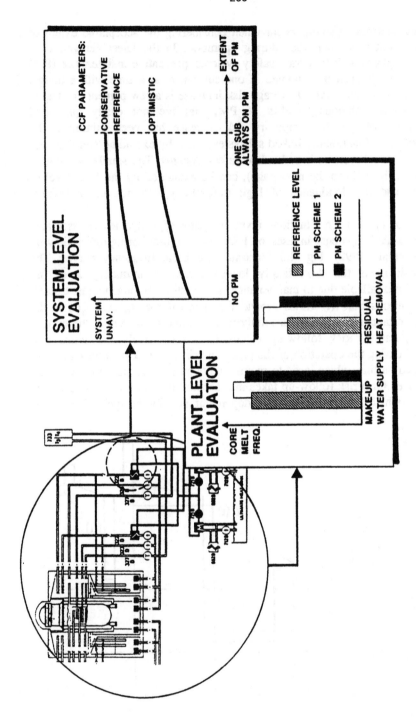

Fig. 4.2. Analyses at system level and plant level of preventive maintenance during power operation at Forsmark 2 (Knochenhauer 1987)

5 Example Accident Sequence - Opportunity for Dynamic Models?

This section describes briefly the most dominant accident sequence in the Forsmark 3 PSA (ABB Atom 1985); the sequence has been a subject of intensive debate and as a follow-up to the PSA some of its special aspects have been addressed in a recent Nordic research project (Hirschberg ed. 1990).

Forsmark 3 is a BWR plant designed by ABB Atom. The plant is characterized by a strict division of all safety systems into four redundant trains which are strictly separated from the physical and functional point of view. Sequence T_fUX2 (T_f = loss of feedwater transient; U = failure of high pressure injection; X2 = failure to initiate manual depressurization) contributes according to the Forsmark 3 PSA 61.4% of the total core damage frequency (CDF) for this plant. The sequence is dominated by:

1. Quadruple CCFs of MOVs in intermittent operation. According to thermal-hydraulic calculations about 50 operations are expected during the mission time of 24 hours.

2. Operator action to initiate manual depressurization. It is worth noting that the corresponding action at elder plants of ABB Atom design is fully automatic. The change to manual actuation in a later design was made in order to minimize the probability of inadvertent depressurization whose consequences were considered as serious (anticipated long investigation period before obtaining permission for continued operation). The time window for initiating the manual action is about 25 minutes. Factors which have been considered as complicating the decision process include: relatively high stress level, possible ambivalence on the side of operators in view of negative consequences of an inadvertent action, and parallel goals of the operators (as illustrated by a simulator exercise operators attempt to recover feedwater supply and high pressure injection). In addition, some training and procedure deficiencies have been identified with regard to the implementation phase.

Some simplifying assumptions in the PSA model are:

— all high pressure injection trains are assumed to fail at time 0;

— recovery of main feedwater and high pressure injection has not been credited;

— when estimating operator failure probability, reactor water level measurement is assumed to be correct.

The recommendation of the Forsmark 3 PSA was to change the logic for the initiation of the depressurization function in order to eliminate the need of manual initiation. The same conclusions were drawn within the Nordic research project (Hirschberg ed. 1990), which included a reference study of the sequence under

208

consideration (Hirschberg et al. 1989). However, the reference study did not include generation of logical models (event tree and fault trees); the models given in the PSA were accepted. Instead, the focus was specifically on the critical elements of the sequence, i.e. the residual CCF contributions for MOVs (previously studied within the CCF data Benchmark Exercise; Hirschberg ed. 1987), the operator failure to initiate manual depressurization (Hirschberg ed. 1989), and treatment of uncertainties. Within the first and third phase of the reference study the participants were free to choose their own approach to the quantification of CCFs and manual depressurization. Phase three represents the best estimate results obtained using experiences from the preceding phases (phase two included comparison of approaches and tools for uncertainty propagation).

Fig. 5.1 shows the survey of reference case results obtained by three groups. The numerical agreement between the groups was drastically improved in the third phase. This is partially due to a more realistic representation of the phased-mission operation of MOVs and use of new data for MOVs within the framework of the linear standby failure rate model.

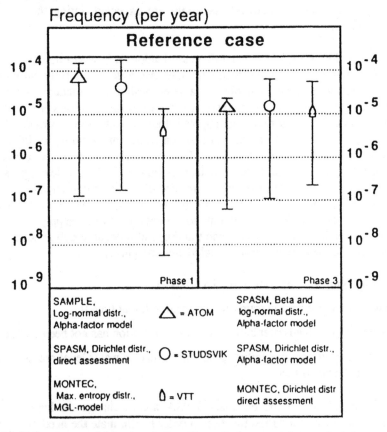

Fig. 5.1. Reference case (phase 1 and phase 3) results for uncertainty analysis for sequence T_fUX2 (Hirschberg et al. 1989)

As a result of recommendations the logic for initiation of manual depressurization was changed at the Oskarshamn plant (a plant belonging to the same generation as Forsmark 3). Furthermore, the corresponding change to fully automatic depressurization was introduced in BWR90, the new BWR design of ABB Atom. As compared to Forsmark 3 PSA the most dominant sequence that includes depressurization was reduced by a factor of 20 and the total CDF by a factor of 4 in the case of BWR90 (Björe et al. 1989). It has to be emphasized that these results were obtained using all the basic assumptions of the Forsmark 3 PSA; some of them were mentioned above. To date no modification of the depressurization logic has been made at Forsmark 3 and the sequences that involve this action are still subject of a debate. The following arguments are used against modification:

— assumption that all trains fail at time 0 is excessively conservative;

— recoveries should have been credited;

— operator action is not a problem given correct water level measurement; the water level measurement as such may be a problem which needs to be addressed in the first place;

— training and procedures have been improved.

As indicated above several time dependencies are important for modeling of the sequence and any improvements in this context could contribute to better understanding, more realistic modeling, and consequently to resolution of the controversy. While some of the problematic points can be resolved employing conventional approaches, modeling of the complex operator interactions involved in the sequence, would probably benefit most from application of dynamic approaches. Thus, the sequence constitutes a challenge for dynamic methods.

6 Conclusions

Generally, treatment of time dependencies is an important but not fully developed topic in current state-of-the-art PSA. The modeling improvements are of large interest and more detailed approaches are becoming necessary in the context of applications of PSAs to support operation and the associated decision-making. The present paper addresses the importance of dynamic approaches in view of existing limitations in the treatment of time dependencies in current PSAs. It is pointed out that dynamic methods have a potential to improve modeling of specific types of time dependency while the limitations associated with other categories will not be reduced. These limitations will be implicitly present also in full scale applications of dynamic approaches. They are mainly related to the existing data sources and to simplifying assumptions that need to be made in the process of data generation. This should not be seen as a criticism of dynamic methods (it would not be reasonable to expect that they would be able to solve all the problems), but rather as an attempt to provide a balanced perspective on the potential capabilities of novel approaches.

The authors believe that the primary problem-solving potential of dynamic methodologies may lie in the possibilities to improve understanding of complex sequences, particularly those including man-machine interactions. This may be more important than the potential benefits of better numerical precision. It is worth noting that there are significant differences between the various dynamic approaches being considered, with respect to their focus; due to their structure some methods are totally focused on quantitative aspects and are consequently less oriented towards generating detailed insights that might help to identify possibly missing scenarios.

It is suggested that some efforts should be made to more clearly define the characteristics of specific situations that may be encountered in PSAs, that are clearly not satisfactorily treated by current simplified approaches but would be by dynamic approaches. The existing evidence is not fully convincing.

From the practitioners' point of view some questions may be asked for consideration by the developers. These questions and authors' comments follow below:

— Which dynamic methods have the best potential to reach sufficient capacity for handling full scale problems? This is a basic requirement and perhaps it might be advisable to compromise the spectrum of applications in order to reach this goal.

— How to assure a proper balance between the advanced character of methodology and data available? The authors' experience is that availability of data rather than lack of methods is in many cases the most limiting factor. Dynamic methods increase the requirements on detailed information and call for additional deterministic analyses.

— Is an extensive decomposition of accident scenarios with respect to time aspects always preferable? To what extent does such a decomposition introduce new uncertainties? One of the experiences from human interactions Benchmark Exercises was that extensive decomposition can lead to overlooking of potentially significant dependencies.

— How big resources are required for application of dynamic approaches in full scale PSAs? Use of such methods definitely leads to needs of additional resources which already are subject to major constraints. It would seem reasonable to apply dynamic approaches, once they have the required capacity, to some selected cases where they can make a difference; for a vast majority of situations current simplified approaches are considered adequate.

References

ABB Atom (1985) Forsmark 3 safety study (in Swedish)

Acosta, C., Siu, N. (1991) Event trees and dynamic event trees: application to steam generator tube rupture accidents. Proceedings of PSA'91 International Symposium on Use of Probabilistic Safety Assessment for Operational Safety, Vienna, Austria, June 3-7, 1991

Björe, S., Hirschberg, S., Öhlin, T. (1989) Probabilistic Safety Analysis for BWR 3300 PSAR. ABB Atom Report RPC 89-54, June 1989

Hirschberg, S. (1990) Impact of parametric and modeling uncertainties and of scope limitations on numerical PSA results. Proceedings of CSNI Workshop on Applications and Limitations of Probabilistic Safety Assessment, Santa Fe, New Mexico, U.S.A., 4-6 September, 1990

Hirschberg, S. (to be published) Prospects for Probabilistic Safety Assessment. Nuclear Safety, 33(3)

Hirschberg, S. ed. (1987) NKA-Project "Risk Analysis" (RAS-470), Summary report on common cause failure benchmark exercise. Report RAS-470(86)14, Nordic Liaison Committee for Atomic Energy, June 1987

Hirschberg, S. ed. (1989) NKA-Project "Risk Analysis" (RAS-470), Summary report on reference study on human interactions. Report RAS-470(89)17, Nordic Liaison Committee for Atomic Energy, December 1989

Hirschberg, S. ed. (1990) Dependencies, human interactions and uncertainties in Probabilistic Safety Assessment. Final report of the NKA Project RAS-470, Nordic Liaison Committee for Atomic Energy, April 1990

Hirschberg, S., Jacobsson, P., Petersen, K.E., Pulkkinen, U., Pörn, K. (1989) A comparative uncertainty and sensitivity analysis of an accident sequence. Proceedings of the Scandinavian Reliability Engineers Symposium 1989, Stavanger, Norway, October 9-11, 1989

Holmberg, J., Johanson, G., Niemelä, I. (1992) Risk measures in Living PSA applications. Report NKS/SIK-1(91)38

Knochenhauer, M. (1987) Plant level probabilistic evaluation of preventive maintenance during power operation in Forsmark 2. ABB Atom Report RPC 87-61, NKA/RAS-450S(87)4, August 1987

Knochenhauer, M. (1988a) Pilot project on valve data analysis. ABB Atom Report RPC 88-59, NKA/RAS-450S(88)3, June 1988

Knochenhauer, M. (1988b) A tentative evaluation of the impact of testing and maintenance on system reliability. Proceedings of the Scandinavian Reliability Engineers Symposium 1988, Västerås, Sweden, October 10-12, 1988

Knochenhauer, M. (1989) Verification of system reliability by analysis of failure data and testing. Proceedings of PSA'89 International Topical Meeting on Probability, Reliability and Safety Assessment, Pittsburgh, Pennsylvania, U.S.A., April 2-7, 1989

Knochenhauer, M., Hirschberg S. (1991) Probabilistically based decision support. Reliability Engineering and System Safety, 36 (1992) 23-28

Laakso, K., Knochenhauer, M., Mankamo, T., Pörn, K. (1990) Optimization of Technical Specifications by use of probabilistic methods, A Nordic perspective. Final report of the NKA project RAS-450, Nordic Liasion Committee for Atomic Energy, May 1990

Mankamo, T., Kosonen, M. (1991) Operational decision alternatives in failure situations of standby safety systems. In: Use of Probabilistic Safety Assessment to evaluate nuclear power plant Technical Specifications, IAEA-TECDOC-599, International Atomic Energy Agency, Vienna, Austria, April 1991

Marseguerra, M., Zio, E. (1992) Approaching dynamic reliability by Monte Carlo simulation. In: T. Aldemir et al. (eds.) Reliability and Safety Assessment of Dynamic Systems. NATO ASI Series F, Vol. 120. Berlin: Springer-Verlag (this volume)

Papazoglou, I. (1992) Markovian reliability analysis of dynamic systems. In: T. Aldemir et al. (eds.) Reliability and Safety Assessment of Dynamic Systems. NATO ASI Series F, Vol. 120. Berlin: Springer-Verlag (this volume)

Vesely, W.E., Hassan, M. (1991) Developments in probabilistic risk evaluation of ageing. Proceedings of the International Conference on Probabilistic Safety Assessment and Management (PSAM), Beverly Hills, California, U.S.A., February 4-7, 1991

Object Representations of Common Reliability Models: New Solution Techniques and Extensions

F. Ann Patterson-Hine and David L. Iverson

NASA Ames Research Center, MS 269-4, Moffett Field, CA 94035-1000, USA

Abstract. Fault tree and digraph models are two common combinatorial models used in dependability analysis of complex systems. An object-oriented representation of the logic structure enables solution algorithms to be developed that take advantage of graph characteristics that are not utilized in previous implementations, enabling more efficient analysis. Algorithms for fault tree quantification, digraph and fault tree cut set enumeration, and dynamic fault tree solution are described. Extensions to standard analysis techniques are enabled with the modular storage of system information associated with the object-oriented approach. These extensions are especially useful for the analysis of large graph structures.

Keywords. Fault trees, digraphs, qualitative analysis, quantitative analysis

1 Introduction

Some of the most challenging problems in the calculation of the reliability of complex systems are determining an efficient information storage and retrieval mechanism for system data, a complementary representation for the reliability model of the system, and algorithms that can integrate and evaluate the system data based on the model structure. Very large complex systems tax current reliability tools because of the number of components that must be modeled resulting in enormous models containing tens of thousands of nodes. Another problem with large systems is the many subsystem and component interactions that exist in them, making both the representation and evaluation of these models very difficult. In addition, more functionality is demanded in tools used by large projects in order to integrate the analysis of reliability with other aspects of system design and operation.

In order to address these issues, reliability models and reliability evaluation algorithms are being developed using the object-oriented programming paradigm. The model is represented by a collection of objects which correspond to component descriptions such as basic events in a fault tree and to the structural aspects of the model such as logic gates in fault trees or states in a Markov chain. The algorithms for evaluating these models are based on conventional algorithms but are redesigned to take advantage of the additional information available in the model description objects. The models developed using this approach are modular and can easily be integrated with other models developed in this environment.

This paper describes the implementation of fault tree and digraph reliability models and evaluation algorithms using object-oriented programming techniques. Section 2 describes the object representation of fault trees and the implementation of an algorithm which quantitatively evaluates the fault tree. Section 3 describes the extension of the object representation to digraph models and the implementations of cut set algorithms for qualitative evaluation of fault trees and digraphs. Section 4 describes the design of an algorithm for evaluating Markov state diagrams based on the object-oriented fault tree techniques.

2 Object-Oriented Fault Tree Algorithms

Fault tree models are a very popular tool in the reliability analysis field. Fault trees can be solved qualitatively, based on a cut set evaluation, and also quantitatively, either by quantifying the cut sets or by direct computation from the fault tree itself. This section describes an algorithm for the direct evaluation of fault trees [1,2]. The direct evaluation is accomplished by an algorithm that combines a simple bottom-up procedure for independent tree branches with a recursive, top-down procedure for branches containing repeated events. The object representation of fault tree events facilitates a unique approach to tree modularization that characterizes tree dependencies dynamically at each step in the reduction process. Results of evaluating intermediate events are stored in event objects as they are calculated. This process results in a reduction of the number of recursive calls required to solve a tree with repeated events and produces the probability of occurrence of all events in the tree.

The following sections describe the fault tree evaluation process. In Section 2.1, object definitions are presented which are used to describe the events in the fault tree model. In Section 2.2, the basic elements of the generalized algorithm are presented in the context of the solution of a very simple fault tree. The tree modularization procedure is discussed in Section 2.3 and the incorporation of the recursive, top-down algorithm is discussed in Section 2.4.

2.1 Fault Tree Object Definitions

Fault trees are made up of events which indicate the propagation of failures throughout a system, starting with one undesirable event, the top event of the tree, and recursively expanding through the causes of that event until component failures are identified as root causes for the occurrence of the top event. Each event in a fault tree can be represented by a data structure called an *object*. Objects are organized into hierarchies, as shown in Fig. 1, with the levels in the hierarchies called *classes*. Component failures are called basic events in a fault tree and are defined out of a class called NODE. Basic events will be referred to as nodes in the remainder of this section. The probability of failure of a component is stored in a named *slot* in each node, called the unavailability slot. In addition to storing the probability of failure of the component, these objects contain pointers to the event's parent(s) in the fault tree in a parent slot.

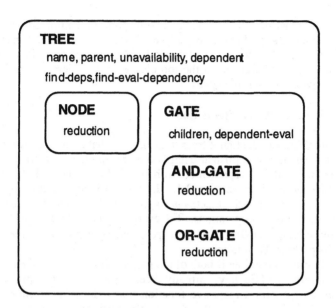

Fig. 1. Object hierarchy.

The top event and all intermediate events in a fault tree correspond to either AND- or OR-logic gates. Logic gates and nodes share many characteristics which can be described by a general class called TREE. Information common to both nodes and logic gates is stored in slots defined in the TREE class. The TREE class is the top level of the object hierarchy. All gates and nodes have both a name and at least one parent, so two slots, name and parent, hold this information. The parent slot described above for the NODE class is actually *inherited* from the TREE class. Since the top event of the fault tree does not have a parent, the value in its parent slot is nil, meaning "empty." Results from the simplification of logic gates are stored in an unavailability slot for gates. This slot is also defined in the TREE class and is where failure data is stored for nodes. Another slot defined in the TREE class is dependent, which indicates whether repeated events are located under a particular gate. Basic events are terminal leaves, making the value of the dependent slot nil for all nodes. The TREE class is specialized into two subclasses, GATE and NODE (defined above), which contain information particular to logic gates and nodes, respectively.

The GATE class includes slots that describe more specifically the state of logic gates. All gates are non-terminal leaves, and the names of their children are stored in a slot called children. A second slot, dependent-eval, indicates whether or not the gate must be evaluated using a top-down algorithm capable of handling repeated events. Specific logic gate types, such as AND- and OR-gates, are described by subclasses that are specializations of the GATE class, AND-GATE and OR-GATE. These subclasses are necessary for defining operations that apply the specific reduction equations required by each type of gate.

Each class has a set of procedures called methods that define the various operations that can be performed on the objects in that class. The evaluation algorithm is modularized into methods that correspond to the procedures that are applied to evaluate the probability of failure of AND-gates and OR-gates. The tree is modularized with another set of methods. As with the data slots described above, methods are generalized in the top classes in the hierarchy and can be inherited and specialized in subclasses.

2.2 Independent Fault Tree Reduction

The reduction of independent fault trees, that is, trees that do not contain repeated events or subtrees, is accomplished using a bottom-up algorithm which is quite simple. The fault tree is represented with the AND-GATE, OR-GATE, and NODE objects defined in Section 2.1. In order to initiate the appropriate evaluation procedures for each object, a *message* must be sent to the object telling it to perform its evaluation operations. The evaluation procedure for the fault tree is initiated by sending a reduction message to the top event of the tree telling it to evaluate its probability of failure.

Evaluation is possible if both of the gate's children are NODEs, or if any children that are GATEs have already been reduced. This procedure is implemented easily with a recursive function that reduces an event's children before attempting to reduce the calling event. The reduction methods for both AND-gates and OR-gates begin by recursively reducing all gates that appear below the particular gate. The next step in the reduction methods differs for each gate type. The reduction method for the AND-GATE class uses Equation 1 to evaluate AND-gates, and the reduction method for the OR-GATE class uses Equation 2 which is appropriate for OR-gates:

$$P(A_1 \text{ AND } A_2) = P(A_1) \cdot P(A_2) \tag{1}$$
$$P(A_1 \text{ OR } A_2) = P(A_1) + P(A_2) - P(A_1) \cdot P(A_2). \tag{2}$$

These equations are appropriate for binary input AND- and OR-gates. Once a gate is evaluated, the result is stored in the unavailability slot, and the procedure returns to the next level of the tree. A gate can be resolved when the failure data of each of its children are available. The reduction method for NODEs is defined for consistency only and does not perform any operations, since NODEs are assigned a probability of failure in the data file. This process continues until the probability of occurrence of the top event is evaluated. Evaluation results in modifications to all of the GATE objects to include the probability of failure of each event determined in the reduction process. The reduction of independent trees with the bottom-up algorithm is very efficient.

2.3 Modularization

Fault trees that contain repeated events are usually modularized before being evaluated to reduce the size and complexity of the tree. Typical modularization procedures identify independent subtrees, reduce the subtrees, and replace them with a node that has a probability of occurrence equal to that of the subtree. Branches that are simplified with these procedures must be independent from every other event in the fault tree. This simpler form of the tree is then evaluated more quickly than the original tree structure. The top-down algorithm which is used here to evaluate dependent trees solves the tree by recursively simplifying groups of events. Therefore, the independence of events should be analyzed with respect to the group which is being simplified and not with respect to the entire tree. The procedure developed to modularize object-oriented fault trees results in a dynamic modularization that analyzes events at each step of the evaluation to determine whether they are independent from the other events in a particular group. This technique is the key factor in improving the performance of the top-down algorithm.

To facilitate modularization, additional information is stored in the tree objects to indicate their dependency status. Two slots, dependent and dependent-eval, were defined for this purpose. The names of all of the repeated events that occur below an event are stored in its object's dependent slot. The names of all of the repeated events that occur more than once below the event are stored in its object's dependent-eval slot. When an object has one or more names stored in dependent-eval, the top-down algorithm must be used to evaluate that event.

The names of repeated events are propagated up the tree by a procedure which examines each node in a recursive, endorder manner. As in the quantification process for independent trees, the children of each node must be examined before the node itself. When a repeated event is encountered, indicated by a list of at least two names in its parent slot, the name of that event is stored in the dependent slot of each of its parents. The names of any events that are stored in the event's dependent slot are also propagated up the tree to its parents' dependent slots. Before a repeated event is propagated to the next level in the tree, it is first examined to determine if the event is already present in the dependent list of the parent event. The names of repeated events are added to this list only if they are not already present. If a repeated event is encountered more than once under a particular event, its name is then stored in the event's dependent-eval slot. When a subtree is repeated and an event inside the subtree is repeated outside the subtree, each event that appears below the top node of the repeated subtree must be added to the subtree top node's dependent list. As each event is added to the list, the names of all of the events that appear below it are also stored in its dependent slot. This ensures proper treatment of events that always occur below the same parent, thus having a single name stored in their parent slots even though they appear in multiple locations in the fault tree. When the examination of an event is completed, all repeated events occurring below that event are listed in dependent, and any event occurring more than once below that event is listed in dependent-eval.

The example tree in Fig. 2 can be used to demonstrate the modularization procedure. Basic events are represented by NODE objects and the logic gates are represented by GATE objects. If a NODE or a GATE represents a repeated event, the parent slot contains a list of the names of the event's multiple parents. Nodes N8 and N9 and gate G6 are repeated events, so their parent slots contain lists of the names of each of their parents.

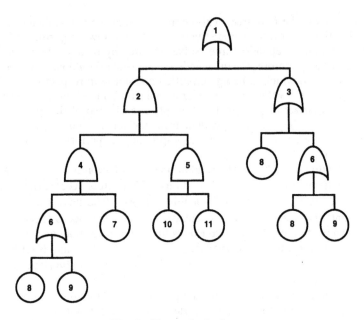

Fig. 2. Example fault tree.

N8 is located beneath gates G6 and G3, so the parent slot for N8 contains the names G6 and G3. N9 is a repeated event; however, it always appears under G6. Thus the parent slot for N9 contains the name of its single parent, G6. The parent slot for G6 contains the names G3 and G4.

The method which defines the examination algorithm is called find-deps. G1 is sent a find-deps message, which initiates the examination process. The children of G1 must be examined before G1 itself, so G2 and G3 are sent find-deps messages, in the order they appear in the children slot of G1. This procedure continues until a basic event is encountered and starts the repeated event propagation process defined in the find-eval-dependency method. N8 is the first NODE to be encountered and it sends itself a find-eval-dependency message. Since N8 is a repeated event, it stores its name in the dependent slots of each of its parents, G6 and G3. N9 is encountered next, but since it always occurs below G6, its parent slot does not contain a list. Therefore, no action is taken. G6, a repeated event, is examined after N9. Since G6 is an intermediate event, and the value of its dependent slot is non-nil, it first stores the names of all events that appear below it in its dependent slot. N9 is added to the list that already contains N8. G6 propagates its name to the next level in the tree, along with N8 and N9 which are in its dependent slot. G6, N8, and N9 are propagated through non-repeated events G4 and G2 to the top event, G1. When G3 is examined, N8 is sent a find-eval-dependency message and stores its name in the dependent slot in G3. The examination of G6, appearing for the second time, results in the propagation of G6, N8, and N9 up to G3 also. Since N8 is already stored in G3's dependent slot, it is stored in the dependent-eval slot. G3 propagates G6, N8, and N9 up to G1. Since they already appear in the dependent slot, these three events are stored in G1's dependent-eval slot.

2.4 Dependent Fault Tree Evaluation Algorithms

Once the fault tree is modularized, the evaluation process begins. The reduction methods used to evaluate independent trees are modified to look for the dependency indicators. Any event whose dependent-eval slot contains nil can be reduced by the simple, bottom-up procedure which uses (1) for AND-GATEs and (2) for OR-GATEs. As objects are evaluated, the results are stored in their unavailability slots.

An event with a non-empty dependent-eval slot must resort to the recursive top-down procedure, PROB-OF-EVENT, for the evaluation of dependent trees. This procedure is patterned after the Page and Perry algorithm [3,4], with modifications to utilize intermediate results and the dependency indicators. The top-down algorithm begins by trying to simplify the event of interest, represented as a set containing one element which is the event of interest. If the event is an AND-gate, the gate is replaced by a set of gates consisting of the children of the AND-gate. The algorithm is then called recursively for the simplification of the new set. If the event is an OR-gate, the algorithm must simplify three new sets of events: one set containing the left-child of the OR-gate, one containing the right-child, and one containing both children. Searching for independence among the events in the set, or among the children of an OR-gate in the set and any other events in the set, can reduce the number of recursive calls considerably. The use of intermediate results to complement these searches reduces the number of recursive calls even more.

When the top-down algorithm is called to simplify a set of events, the set is initially searched to determine if any events are independent from every other event in the set. If such an event is found, it is marked so that it is not simplified further. It effectively becomes a NODE, since its probability value is stored from a previous calculation. This corresponds to the following equation used by Page and Perry [3]:

$$\text{PROB-OF-EVENT}(S) = \text{PROB-OF-EVENT}(\{n\})$$
$$\cdot \text{PROB-OF-EVENT}(S - \{n\}) \quad (3)$$

The use of the intermediate results, however, eliminates the need to solve the first term, PROB-OF-EVENT($\{n\}$). When all independent events in the set are marked, the set is then searched for AND-GATEs. Any AND-GATEs appearing in the set are replaced by their children. The final search for independence involves searching the remaining OR-GATEs for independence in their children from the other events in the set. If an independent child is found, then (4) is used to simplify the set:

$$\text{PROB-OF-EVENT(S)} = \text{PROB-OF-EVENT}(\{j\} \cup S - \{n\})$$
$$+ [1 - \text{PROB-OF-EVENT}(\{j\})]$$
$$\cdot \text{PROB-OF-EVENT}(S \cup \{k\} - \{n\}) \quad (4)$$

with j being the independent child of node n, and k the dependent child. In the present case, however, the solution of PROB-OF-EVENT($\{j\}$) is eliminated since its probability value is known from a previous calculation.

3 Object-Oriented Cut Set Algorithms

Qualitative evaluation of combinatorial reliability models, such as fault trees and digraphs, is a common operation. One popular qualitative evaluation technique is enumeration of minimal cut sets. Cut sets are sets of basic failure events that could cause a specified high level failure. Minimal cut sets are cut sets with the property that if any one of the failure events is removed from the set, the remaining failures would not be sufficient to cause the high level failure for which the cut set was calculated. The object-oriented programming paradigm is well suited to the task of calculating minimal cut sets of digraph and fault tree models.

The following sections describe object-oriented minimal cut set algorithms for fault tree and digraph models. Section 3.1 discusses a fault tree cut set algorithm that uses an object-oriented fault tree representation similar to that described in Section 2.1. In Section 3.2, matrix based digraph cut set algorithms are discussed briefly, then a technique to extend the fault tree cut sets algorithm to solve digraph models is explained.

3.1 Fault Tree Cut Set Calculation

As discussed previously, fault trees can be represented as a set of AND-GATE, OR-GATE, and basic event NODE objects. The structure of the tree is stored in lists of object pointers in the parent and children slots of each object. The object-oriented fault tree cut set algorithm requires only one additional slot in each object. This slot is called the cutsets slot, and it is used to store calculated cut sets for each gate in the fault tree.

A method called find-cutsets is defined for each class of object in the fault tree. The find-cutsets method for a basic event NODE object simply returns a one element singleton set containing that basic event. Find-cutsets for an OR-GATE object returns all the cut sets of the children of that OR-gate. The AND-GATE find-cutsets method takes the cross product of the cut sets of each of the AND-gate's children. For instance, if an AND-gate has three children, the method will form one cut set with the union of the first child's first cut set, the second child's first cut set, and the third child's first cut set. A second cut set will be formed with the union of the first child's first cut set, the second child's first cut set, and the third child's second cut set. This continues until all possible combinations have been formed. When solving for minimal cut sets, the find-cutsets methods for AND-gates and OR-gates will perform one additional step. After gathering the cut sets from the gate's children and combining them appropriately, every set in the group of cut sets that is a superset of another set in the group is removed. This process leaves only minimal cut sets. Cut sets that are larger than the maximum desired size can also be removed at this point.

The calculation of cut sets is a recursive process. The cut sets of a fault tree gate are built from the cut sets of its children. The find-cutsets methods used in the object-oriented cut set algorithm are recursive functions. When a GATE object receives a find-cutsets message, it sends a find-cutsets message to each of its children to obtain their cut sets. Once the cut sets are calculated for a given object, they are stored in the cutsets slot of that object. This prevents recalculation of the gate's cut sets every time the object receives the find-cutsets message. Initially the cutsets slots in each object are empty. When an object receives a find-cutsets message it first checks to see if any cut sets are stored in its cutsets slot. If the slot is empty, find-cutsets messages are sent to each child and the gate's cut sets will be calculated. If the slot contains cut sets, no recalculation is required, the method just returns the cut sets found in the cutsets slot. This saves a significant amount of computation on fault trees containing repeated events since a repeated event's cut sets are calculated only once instead of being recalculated for each time the event appears in the tree. In addition, when a find-cutsets message is sent to the root of a fault tree, you not only find cut sets for the root, you will get cut sets for every gate in the tree.

3.2 Digraph Cut Set Calculation

Although the modeling technique differs, digraph models are very similar to fault tree models when both are considered as Boolean graph models. Both are failure space models. Digraphs consist of AND-gates and nodes connected by directed edges. Nodes in the digraph behave like fault tree OR-gates. Fault trees have AND-gates and OR-gates with an implied directional flow from bottom to top. Each node in a digraph model represents a failure in the system. In a fault tree each basic event node and each gate represent a failure. The main differences between digraphs and fault trees are that digraphs have no designated basic event nodes, digraph models allow loops or cycles in their structure, and digraphs express information about propagation of failures with the directed edges. The cut set definitions differ slightly for each model. The cut sets for a fault tree gate are sets of basic failure events that could combine to cause the failure represented by that gate. The cut sets for a digraph node are sets of digraph nodes that, when the failures they represent occur concurrently, will cause the failure represented by that digraph node to occur.

Most digraph cut set solution codes represent the digraph model with matrices. A digraph containing N nodes is represented by two NxN matrices. One matrix shows connectivity between nodes by placing a one in the [X,Y] cell of the matrix to indicate that node X is connected to node Y. The other matrix shows which nodes combine as inputs to AND-gates. An AND gate with input nodes A and B and output node C would be represented by ones in cells [A,C] and [B,C]. To find digraph cut sets using a matrix representation, several matrix calculations involving logical operations between a vector and each row in the matrix must be computed for every node in the digraph. This procedure can be computationally costly. Additionally, many of these algorithms are limited to singleton and doubleton solutions due to their computational complexity.

An object-oriented approach simplifies the digraph cut set solution problem and allows cut sets of any size to be calculated. The object-oriented digraph representation is similar to the fault tree object representation. Each digraph node and each AND-gate are represented by an object. Instead of parent and children slots, the digraph objects have input and output slots listing their input nodes and output nodes respectively. The digraph objects also have a cutsets slot to store calculated cut sets, and a visited slot that is used to keep the algorithm from entering infinite loops. The object-oriented algorithm used for digraph cut set calculation is a modified version of the fault tree algorithm. The cut sets for each digraph node consist of a singleton representing that node plus the cut sets from each of the node's inputs. The cut sets for a digraph AND-gate are the cross product of the cut sets of its inputs. The major difference between the digraph and fault tree cut sets algorithms is that the digraph algorithm must keep track of the nodes it has visited during the solution process so it will not enter an infinite loop when it solves for nodes in a cycle.

A find-cutsets method is defined for each class of digraph object. When an object receives the find-cutsets message, the first thing it does is check if any cut sets are stored in its cutsets slot. If so, it returns those cut sets. If the cutsets slot is empty, it checks if its visited slot is marked. If the visited slot is marked, the node or AND-gate it represents is contained in a loop and has already been visited by the solution algorithm so any cut sets that are calculated from its input nodes will be repeats of cut sets already found. In this case the find-cutsets method returns nothing. This also prevents the algorithm from going on to process the node's inputs and getting caught in an infinite loop. If the visited slot of an object receiving the find-cutsets message is not marked when a find-cutsets message is received, the slot is marked and the cut sets of each input node or AND-gate will be found by sending find-cutsets messages to the objects listed in the inputs slot. The resulting cut sets will be combined in a manner appropriate for the object class (see fault tree find-cutsets description for details), and if the object represents a digraph node, a singleton cut set containing that node is added. At this point non-minimal cut sets and cut sets larger than a specified maximum size can be removed. Then the visited slot of the object is unmarked, and the calculated cut sets are returned in response to the find-cutsets message. If no loops were encountered in the cut set solution process or if this node received the original find-cutsets message, the cut sets will be stored in the object's cutsets slot. If this was not the originating node and a loop was encountered, the cut sets are not stored since they will not be a complete cut set solution for that node.

The advantages of the object-oriented digraph cut sets algorithm are the ability to store intermediate results, the straightforward representation of graph connectivity, and the ability to solve for cut sets of any size. As with the fault tree cut sets algorithm, storing cut set results in the digraph objects during the course of the solution can eliminate a significant amount or recomputation. Finding the inputs and outputs of nodes and AND-gates in the object-oriented digraph representation is done by simply following a memory pointer to another object from the lists of object pointers in an object's inputs and outputs slots. This is usually more efficient than the matrix operations required with the matrix representation. Finally, this algorithm can solve for any size digraph cut set, unlike many matrix based digraph codes that are limited to singletons and doubletons.

4 Extensions to Dynamic Modeling

The dynamic behavior of systems cannot be captured in static models such as fault trees and digraphs. Past work in the area of fault tolerant computer system modeling has resulted in a technique which combines the power of Markov chains to capture dynamic system behavior such as transient errors, fault recovery, and sequence dependent failures and a top level fault tree representation. The fault trees are automatically converted to a Markov chain, including the necessary dynamic behavior, and then the resulting Markov chain is solved as a set of linear ordinary differential equations [5,6]. In this way, the advantages of both model types, fault trees and Markov chains, can be exploited. New gate types were defined that capture sequence dependent behavior, such as cold spares and sequence dependencies [7]. These new gates were necessary for representing the behavior of many fault tolerant computer systems in dependability models. These models are called dynamic fault trees.

A modular approach for solving dynamic fault trees draws heavily on the representation and solution methods of fault trees using object-oriented programming techniques. This process is explained in detail in [10]. The modularization process described above for static fault trees is extended to dynamic fault trees. The identification of independent subtrees is a key to the development of hierarchical models. Once independent subtrees are identified, the most appropriate solution technique can be applied to each independent submodel, and then the results of the submodels can be combined into higher level fault trees. Once the submodels are identified, a solution technique is chosen ranging from standard combinatorial solution to full Markov solution for each submodel. Hence, different solution techniques can be used for each submodel. This approach reduces the size of the state space needed to represent the dynamic portions of the fault tree model, enabling more accurate solutions of models for long mission scenarios.

A Fault Tolerant Parallel Processor (FTPP) cluster [8,9] is used to illustrate application of the above techniques. The cluster consists of sixteen processing elements (PE), with four connected to each of four network elements (NE). The network elements are fully connected and form a Byzantine resilient core for the cluster. Four of the processors (one on each NE for Byzantine resilience) form a quad redundant processing group. Three other processors form a triad, while the remainder of the processors are used as hot spares. The simplex elements on a NE can spare for any failed PE on the same NE.

A fault tree was constructed for the scenario that system failure occurs if either the quad or the triad processing groups fail. The Markov model that resulted from the fault tree conversion had 1245 states and 5847 transitions, and required approximately 9 CPU minutes to generate on a DECstation 3100. The

Markov chain was truncated after the consideration of three component failures and required approximately 99 CPU minutes to solve for both phases. This solution was good for short to medium length phases (up to a week long), but produced unacceptably wide error bounds for longer missions. Models for longer missions required more states, as it was not reasonable to truncate the model at such a low level since multiple failed elements were more likely with long missions. Unfortunately, the size of the model doubled for each additional failure level considered.

Instead of generating a larger Markov model, the fault tree was solved by decomposing it into several independent subtrees. These subtrees were identified through the use of object oriented techniques, since the objects (gates) store information about their descendants. Once the independent subtrees were identified, the Markov model that was required to represent the system's dynamic behavior was reduced to 18 states and 34 transitions. An exact solution of this model was easily obtained, therefore, the model was applicable for long mission scenarios as well as short ones. This example is explained in more detail in [10].

References

1. Patterson-Hine, F.A.: Object-Oriented Programming Applied to the Evaluation of Reliability Fault Trees. Ph.D. Dissertation. The University of Texas at Austin, Mechanical Engineering Department (1988).
2. Patterson-Hine, F.A. and Koen, B.V.: Direct Evaluation of Fault Trees Using Object-Oriented Programming Techniques. IEEE Transactions on Reliability, June (1989).
3. Page, L. B., and Perry, J. E.: An Algorithm for Exact Fault-Tree Probabilities without Cut Sets. IEEE Transactions on Reliability R-35, 544-558 (1986).
4. Page, L. B., and Perry, J. E.: A Simple Approach to Fault-Tree Probabilities, Computers and Chemical Engineering 10, 249-257 (1986).
5. Dugan, J. B., Trivedi, K. S., Smotherman, M. K. and Geist, R. M.: The Hybrid Automated Reliability Predictor. AIAA Journal of Guidance, Control and Dynamics, May-June (1986).
6. Boyd, M. A.: Converting fault trees to Markov chains for reliability prediction. Master's thesis. Duke University, Department of Computer Science (1986).
7. Dugan, J. B., Boyd, M. and Bavuso, S.: Fault trees and sequence dependencies. Proceedings 36th Annual Reliability and Maintainability Symposium (1990).
8. Harper, R. E.: Reliability Analysis of Parallel Processing Systems. Proceedings of the 8th Digital Avionics Systems Conference, October (1988).
9. Harper, R. E., Lala, J. H. and Deyst, J. J.: Fault Tolerant Parallel Processor Architecture Overview. Proceedings of the 18th Symposium on Fault Tolerant Computing, June (1988).
10. Patterson-Hine, F.A. and Dugan, J. B.: "Modular Techniques for Dynamic Fault Tree Analysis, 1992 Annual Reliability and Maintainability Symposium, Las Vegas, NV, January (1992).

The Approach of a Regulatory Body to the Assessment of Risks from Hazardous Installations

Sam Porter

Major Hazards Assessment Unit, Health and Safety Executive, St. Annes House, Bootle, Merseyside, L20 3RA, U.K.

Abstract. This paper describes how the theme of this workshop can be viewed in the context of a real world example of decision making based on quantified risk assessment.

Keywords: quantified risk assessment; land use control; hazardous installations; Regulatory body.

1 Introduction

The Health and Safety Executive (HSE) is the Regulatory Body in the UK responsible for monitoring the compliance of industry with relevant health and safety legislation. Additional control measures are adopted for 'hazardous installations' where an accident on site has the potential to cause significant off-site consequences due to, for example, fire, explosion or the release of a toxic gas cloud (hazardous installations discussed in this paper exclude nuclear, offshore, mines and certain other installations). In addition to further Regulations covering on-site safety, the HSE also gives advice on land use in the vicinity of hazardous installations. This paper discusses how quantified risk assessment is used in decision making about development proposals (e.g. residential, shopping, etc.) which could be affected in the event of a major accident at the hazardous installation.

2 The Control of Hazardous Installations

The UK approach to dealing with hazardous installations is essentially a 3 point strategy [1, 2, 3, 4, 5].

---

1   The views expressed herein are those of the author and do not necessarily reflect the views of the Health and Safety Executive.

2.1 Identification

In any system of control for major hazards, the first step is to establish suitable criteria to recognise those installations which present the greatest potential threat to safety in terms of their potential to cause serious off-site consequences in the event of a major accident involving loss of containment. Such criteria have been chosen/developed in terms of quantity of hazardous substance and the nature and severity of the hazard. This information has been embodied in Regulations as a list of named substances and generic categories which can be used to identify hazardous installations.

2.2 Prevention

Compliance with all relevant health and safety legislation should minimise the likelihood of an occurrence of a major accident. The Control of Industrial Major Accident Hazards Regulations 1984 (CIMAH) implement the European Communities' 'Seveso' Directive. Their aim is to prevent major chemical industrial accidents and to limit the consequences to people and the environment of any which do occur. A basic requirement is that a company in control of an industrial activity can demonstrate that potential causes of a major accident have been identified and that adequate steps have been taken to prevent such accidents and limit their consequences. Additional more stringent requirements apply to the potentially more hazardous activities. These include duties on the company to submit a written safety report to the HSE, prepare an on-site emergency plan, and provide certain information to the public. The Local Authority is required to prepare an off-site emergency plan based on information provided by the company.

2.3 Mitigation

In general, the attainment of absolute safety is not considered possible and it is therefore considered prudent to take account of the possibility of a major accident. Where there is potential for such an accident, mitigating measures can be taken to reduce the impact upon people outside the installation. Firstly, planning law can be used to keep new major hazard plants away from centres of population wherever possible; similarly, it is considered wise to use planning law to avoid a substantial growth in population near an existing installation. Secondly, the effects of any major accidents which do occur can be lessened by putting into action previously prepared emergency plans. These should enable companies, local authorities and the emergency services to cooperate efficiently for the protection of the public and the environment.

2.4 Basic Principles

The basic principles of major accident prevention/mitigation may be summarised as follows:

- *Design* to adequate standards
- *Construct* to adequate standards
- *Commission* carefully, with appropriate feedback
- *Modify* carefully, with adequate modification procedures in place and re-assessment of safety as necessary
- *Operate* correctly
- *Respond* correctly to plant deviations
- *Maintain/test* properly
- *Limit* the number of *people* at risk
- *Plan for emergency* to reduce effects of accident

3 The Use of Quantified Risk Assessment

Quantified risk assessment (QRA) is used in two main areas within the framework discussed above for controlling the major accident potential of a hazardous installation.

3.1 Safety Reports

A safety report is required to provide information about the dangerous substance, the installation, the management system and describe potential major accidents and the measures taken to prevent, control or minimise the consequences of any such accident. QRA is often used as a technique to investigate the adequacy of existing preventative/control measures and may often support an assertion that additional plant modifications are not required.

3.2 Land Use Control

It may readily be appreciated that the risk of death or serious injury from a major accident at a hazardous installation reduces as the distance from the site becomes greater. The HSE will normally undertake a plant hazard/risk assessment to establish how close a development may be allowed to such an installation or vice versa. The predicted level of risk at a particular location is used to form the basis for advice HSE will give to the Local Planning Authority in accordance with adopted criteria for 'tolerable risk levels' [1]. The setting of such criteria is beyond the scope of this paper.

A typical risk assessment will consider postulated releases from all major storage/process plant and associated pipework, and investigate the consequences of such releases. This includes assessment of:

- frequency of events involving loss of containment;
- fluid release rates and duration of releases, ranging from holes and pipework breaks to complete vessel failure;
- dispersion of flammable or toxic gases, taking into account different types of weather and the probability of the wind blowing in a particular direction;
- the likelihood of serious injury from a particular hazard; this may include consideration of the probability of injury from predicted levels of:
  - blast over-pressure from an explosion;
  - thermal radiation from a fire;
  - concentration of a toxic gas.

The above quantification is inevitably subject to considerable uncertainty. The nature and degree of uncertainty varies with the style of QRA which is undertaken, and with the hazard being assessed [3]. The two main types of assessment are
- the use of 'conservative/pessimistic' inputs and assumptions;
- the use of 'best estimate' inputs and assumptions.

HSE currently uses an approach which may be described as 'cautious best estimate'. It is intended that assumptions and inputs are realistic, but some over-estimate is preferred when doubt arises. There is clearly a need to fit the judgement criteria to the assessment approach. The criteria for tolerable risk have been set quite low for this best estimate approach, to make some allowance for the less likely occurrence of an event which compounds pessimisms.

The end result is a 'cautious best estimate' prediction of the probability of persons receiving serious injury at a certain distance from the plant. This information is then used to make a decision on proposed developments in accordance with adopted criteria for tolerable risk [1].

4 Discussion/Conclusions

In the context of this particular workshop, this paper was presented in the session 'Risk Assessment for Realistic Decision Making' as an example where QRA is regularly used in judging relative safety in order to advise on proposals for development in the vicinity of hazardous installations.

The theme of the workshop was directed at investigating the need for improved methods for dealing with 'dynamic systems', incorporating the time dependency of plant configuration and the operator. However the degree of imperfection of existing simpler methods was unclear.

An enhanced approach for dynamic plant systems would largely affect only one of the inputs (event probabilities) to the risk assessment as discussed in Section 3.2. Small changes to a 'best estimate' of this single variable may not necessarily have a significant effect on the resulting decision.

It may be noted that the many inputs to a QRA can typically give rise to an 'uncertainty' equivalent to 2-3 orders of magnitude difference in predicted risk levels. A 'best estimate' approach combined with appropriate judgement criteria can help deal with this uncertainty, but the approach remains fairly crude. Furthermore QRA cannot be expected to adequately cover all of the basic prevention areas highlighted in Section 2.4. Against this background, the use of a consistent approach for assigning values for input variables and judgement criteria can often be more important than refinement of one particular input.

Despite the above cautionary note, the improvement of risk assessment methods is obviously desirable. However it is important to clearly define when enhanced analysis techniques are considered necessary. The HSE receive over 400 safety reports from companies operating widely different plant and processes, varying from relatively simple to very complex. It is therefore necessary, from the point of view of a Regulatory Body, to have a clear definition of the areas where existing methods may be inadequate and ideally some 'feel' for the extent of such inadequacy.

References

1. Health and Safety Executive: Risk criteria for land-use planning in the vicinity of major industrial hazards. Her Majesty's Stationary Office, London.
2. Health and Safety Commission: The control of major hazards, third report by the Advisory Committee on Major Hazards. Her Majesty's Stationary Office, London.
3. Health and Safety Executive: Quantified risk assessment: its input to decision making. Her Majesty's Stationary Office, London.
4. Health and Safety Executive: A guide to the Control of Industrial Major Accident Hazards Regulations 1984. Her Majesty's Stationary Office, London.
5. Health and Safety Executive: Tolerability of risk from nuclear power stations. Her Majesty's Stationary Office, London.

Part 5

Working Sessions

Summary of Working Session Activities

Nathan. O. Siu,[1] Tunc Aldemir,[2] Ali Mosleh,[3] P. Carlo Cacciabue,[4] B. Gül Göktepe[5]

[1] Center for Reliability and Risk Assessment, Idaho National Engineering Laboratory, Idaho Falls, ID 83415, USA
[2] The Ohio State University, Nuclear Engineering Program, Columbus OH 43210, USA
[3] Materials and Nuclear Engineering Department, University of Maryland, College Park, MD 20742, USA
[4] Institute for Systems Engineering and Informatics, Commission of the European Communities Joint Research Center, 21020 Ispra (Varese), ITALY
[5] Çekmece Nuclear Research and Training Center, Istanbul, TURKEY

1 Introduction

As discussed in the preface to these proceedings, the primary objectives of the ARW were to: a) discuss the advantages and disadvantages of proposed methodologies for analyzing the reliability and safety of dynamic process systems, b) identify practical situations where these methodologies could result in significantly improved results, and, c) develop a benchmark exercise to compare the dynamic methodologies with each other and with conventional event tree/fault tree analysis. The structure of the ARW was constructed to accomplish these objectives through a set of formal paper presentation sessions and a set of working sessions. The former set of sessions was designed to open discussion on the proposed methodologies for analyzing the reliability and safety of dynamic process systems and the latter set was designed to review and discuss these methodologies and to directly define a benchmark exercise.

During the course of the ARW, it became apparent that additional time was needed to discuss a variety of issues, including the motivation for dynamic analyses, the desired technical characteristics of the benchmark exercise (e.g., problem complexity) and practical aspects of the exercise (e.g., approach to obtain funding). The working session structure was therefore modified to address these issues.

This summary outlines the original working session structure, the modified working session structure and the results of the working sessions. It is shown that, despite the reduced amount of time available for defining the benchmark exercise, considerable progress was made in this direction. In particular, the general characteristics of a standard problem to be used in the benchmark exercise were specified and a project plan to fund and execute the exercise was developed.

2 Original Structure

The original structure of the working session portion of the ARW included four major working sessions, arranged in two sets of two parallel sessions, and a number of summary sessions following these working sessions. Table 1 shows the organization of the technical portion of the ARW as envisioned at the beginning of the meeting.

Table 1. The Original ARW Structure (Technical Portion)

Session Type (Date)	Title/Scope (Session Number)
Formal Presentations (8/25/92, 8/26/92)	• Dynamic Approaches - Issues and Methods (P1) • Dynamic Approaches - Applications (P2) • Human Reliability and Dynamic Systems Analysis (P3) • Risk Assessment for Realistic Decision Making (P4)
Group Discussion (8/27/92 a.m.)	Review of session summaries, working group objectives and structure (R0)
Parallel Working Sessions (8/27/92 a.m.)	• Dynamic PRA Capabilities: Separating Hope From Hype (W1) • The Real World: What Should Dynamic PRA Address? (W2)
Group Discussion (8/27/92 p.m.)	Review of the working sessions W1 and W2 (R1)
Parallel Working Sessions (8/27/92 p.m.)	• Benchmark Problem Definition A (W3) • Benchmark Problem Definition B (W4)
Group Discussion (8/28/92 a.m.)	Review of the working Sessions W3 and W4 (R2)
Group Discussion (8/28/92 a.m.)	Workshop evaluation and future plans (R3)

The first two parallel working sessions (i.e. W1 and W2) were aimed at: a) realistically assessing the current capabilities of the dynamic methodologies discussed in the formal presentations, and, b) identifying the desired characteristics of an ideal methodology as demanded by real-world problem solving needs. The sessions were to be chaired by the ARW organizers. In both cases, participants were asked to develop a hierarchical structure to represent the model characteristics, actual in session W1 and desired in session W2.

The second two sessions (i.e. W3 and W4) were aimed at developing suitable benchmark problems, based on the results of sessions W1 and W2. Naturally, the benchmark problems were required to be sufficiently complex to allow a comparative evaluation of the various dynamic methodologies applied towards their solution. Furthermore, they were required to be realistic and relevant to current safety concerns, in order that they be of interest to outside groups.

These requirements provided a considerable driving force in the modification of the working session structure during the meeting, as discussed in the following section.

3 Modified Structure

During the summary session R0 prior to the start of the working sessions (see Table 1), considerable discussion arose as to the exact intent of the working sessions and the degree to which practical issues in conducting a benchmark exercise (e.g., funding of participants) had been addressed. It became apparent that the proposed working session structure required modification in order to adequately address these issues.

With the agreement of the ARW participants, two working groups were formed to redirect the objectives of working sessions W1 - W4. Working Group 1 was asked to define "dynamic systems" in such a way that differences between the methodologies discussed in the formal presentations and conventional methodologies could be more clearly defined and to discuss a number of important issues related to the conduct of a benchmark exercise for dynamic methodologies (e.g., performance measures for the different methodologies). Working Group 2 was tasked with developing boundary conditions for the exercise and a preliminary strategy for actually conducting the exercise. To allow redundant coverage of key issues, some of the specific discussion items were addressed by both groups. As with the originally planned structure, the results of the working group sessions were discussed in a joint session by all workshop participants (i.e session R2).

Additional details on the individual working group assignments are provided in Table 2. The results of the working group discussions are summarized in the following section.

Table 2. Tasks Assigned to Working Groups in the Modified Working Session Structure

Working Group Number	Tasks
1	• Definition of dynamic systems
	• Motivation for dynamic analyses
	• Methodology related issues
	- Objective of the bechmark exercise
	- Figures of merit for comparing methodologies
	- Sensitivity analyses
	- Specifictions for future developments
	• Funding for exercise
	- Form/approach
	- Sources
	- Problem/funding coupling
2	• Definition of dynamic systems
	• Complexity of system treated in benchmark exercise
	- Physical processes
	- Automatic actions/human actions
	- Number of components
	- Branching
	- Stochastic aspects
	• Structure of the benchmark exercise/alternatives
	• Funding for exercise
	- Form/approach
	- Sources
	- Problem/funding coupling

4 Working Session Results

The working group sessions and the ensuing session discussing the working group results were fruitful. It was pointed out that while all systems are dynamic (i.e. their behavior and governing parameters change over time) the need for special dynamic analysis methodologies is determined by the time constants of change. This is an important consideration when selecting a problem for comparing dynamic and conventional methodologies. It was also recognized that the term "benchmark problem" implies the existence of a correct answer against which all candidate methodology predictions would be compared. In practical situations, such a correct answer may either not exist or may be too difficult to compute; the term "standard problem" was therefore adopted to represent the problem to be used when comparing different methodologies.

More importantly, the working sessions resulted in a number of boundary conditions for the standard problem and in a practical plan for conducting a benchmark exercise to compare proposed and standard methodologies on this problem. Regarding the standard problem, it was resolved that the standard problem:

- should be constructed to allow the evaluation of the ability of both the dynamic and static (i.e. conventional) methodologies to provide a realistic assessment of the selected performance measure (e.g., risk),

- must be sufficiently complex to challenge the dynamic methodologies and provide aspects that demonstrate the differences between dynamic and static approaches, but,

- must be simple enough to allow an exact solution (closed form or numerical) for all or part of the system behavior and allow the simplifications or approximations that are required in order to apply traditional analysis methods.

A sample standard problem with these characteristics was suggested. The sample problem involved a storage tank similar to that discussed in [1] and later exercised in [2] and [3]. Like the earlier problem, the suggested standard problem involved a mixture of logical variables (for hardware state) and continuous variables (for processes), automatic controllers and stochastic failures. Unlike the earlier problem, the suggested problem involved multiple process variables (instead of one) and complex, non-linear laws of behavior. It was pointed out that not only does the suggested problem have characteristics of interest to a benchmarking exercise, it also has close structural similarity to a number of real problems (e.g., aircraft flight controller reliability, nuclear power plant shutdown risk, nuclear waste storage tank criticality risk).

Regarding the plan for conducting benchmark exercise, two potentially critical problems were identified. First, it was recognized that funding for participants could be difficult to obtain. Second, it was observed that due to the relative immaturity of available human reliability analysis methodologies, it would be hard to specify a standard problem involving general forms of human error that tested the capabilities of methodologies, rather than those of the analysts exercising the methodologies.

To address the first problem, it was assumed that while potential sponsors would be unlikely to fund (at a sufficiently high level) benchmark exercise activities for their own sake, they might be more interested if the solution of the standard problem is a useful step in solving problems of direct interest to them. It was therefore decided to employ a standard problem that could be extended with a little effort to address current safety concerns in a variety of industries. The suggested sample problem described above appears to have just this characteristic. Furthermore, it was decided to initiate work on a standard proposal that would focus on the solution of the standard problem using dynamic analysis, but could also be easily modified to address problems of more direct interest to a potential sponsor.

To address the second problem, it was decided to conduct the benchmark exercise in a phased manner. In the early portion of the exercise, only limited human actions (essentially those for which the human can be treated as procedure-following automaton) would be treated. In a later portion, more general behaviors would be addressed (presumably with models currently being developed).

As a final result of the working session portion of the advanced research workshop, three committees were established to continue formulation of the benchmark exercise program. The main committee was tasked with further developing the plan for the benchmark exercise, and with developing the basis for a standard proposal mentioned above. The members of this main committee are as follows (see Appendix A-List of Participants for affiliations and addresses):

- P. C. Cacciabue (European coordinator)

- N. O. Siu (U.S. coordinator)

- T. Aldemir

- B. G. Göktepe

- A. Mosleh

- P. Wieringa

The remaining two committees were tasked with supporting the work of the main committee. The first support committee was assigned with the detailed development of the standard problem. Its members are:

- T. Aldemir

- B. Walker

- J. Wreathall

The second support committee was asked to develop information needed to extend the standard problem to problems of direct interest to particular agencies. Its members are:

- D. Bley (for waste tank applications)

- P. Kafka (for nuclear power plant shutdown applications)

It is anticipated that the work of these three committees will be completed early in 1993.

References

1. Aldemir, T.: Computer-Assisted Markov Failure Modeling of Process Control Systems. IEEE Transactions on Reliability, R-36, 133-144 (1987).

2. Deoss, D. L. and Siu N.: A Simulation Model for Dynamic System Availability Analysis. MITNE-287, M.I.T. Department of Nuclear Engineering, October 1989.

3. Marseguerra, M., Zio E.: Approaching Dynamic Reliability By Monte Carlo Simulation. These proceedings.

List of Participants

Prof. **Tunc Aldemir**
The Ohio State University
Nuclear Engineering Program
206 West 18th Avenue
Columbus OH 43210
U.S.A.

Dr. **Ulvi Adalıoğlu**
Cekmece Nuclear Research and
Training Center
Havaalanı, P.K. 1
İstanbul
TURKEY

Prof. **George Apostolakis**
University of California
Mechanical, Aerospace and
Nuclear Engineering Department
School of Engineering and
Applied Science
38-137 Engineering IV
Los Angeles, CA 90024-1597
U.S.A.

Mr. **Nazım Bayraktar**
Turkish Atomic Energy Authority
ANAEM, Fen Fakultesi Arkası
Beşevler, Ankara 06100
TURKEY

Dr. **Dennis C. Bley**
PLG, Inc.
4590 Macarthur Blvd., Suite 400
Newport Beach, CA 92660-2027
U.S.A.

Dr. **P. Carlo Cacciabue**
Commission of the European
Communities Joint Research Center
Institute for Systems Engineering and
Informatics
21020 Ispra (Varese)
ITALY

Dr. **Giacomo Cojazzi**
Commission of the European
Communities Joint Research Center
Institute for Systems Engineering and
Informatics
21020 Ispra (Varese)
ITALY

Prof. **Jaques Devooght**
Universite Libre de Bruxelles
Faculte des Sciences Appliques
Service Metrologie Nucleaire (CP 165)
Avenue F. D. Roosevelt 50
B-1050 Brussels
BELGIUM

Prof. **Balbir S. Dhillon**
University of Ottawa
Engineering Management Program
Ottawa, Ontario K1N 6N5
CANADA

Dr. **David Gertman**
Idaho National Engineering
Laboratory
Human Factors and Systems
Analysis Unit, MS 2405
P. O. Box 1625
Idaho Falls, ID 83415
U.S.A.

Mrs. **B. Gül Göktepe**
Çekmece Nuclear Research and
Training Center
Havaalanı, P.K. 1
İstanbul
TURKEY

Dr. **Stefan Hirschberg**
Paul Scherrer Institute
Systems/Safety Analysis Section
CH-5232 Villigen PSI
SWITZERLAND

Dr. **Erik Hollnagel**
Human Reliability Associates, Ltd.
School House, Higher Lane, Dalton
Parbold
Lancs. WN 8RP
U.K.

Dr. **Frank R. Hubbard**
FRH Inc.
P. O. Box 65359
Baltimore, MD 21209
U.S.A.

Dr. **Jose M. Izquierdo**
Consejo de Seguridad Nuclear
Justo Dorado 11
Madrid 28020
SPAIN

Dr. **Peter Kafka**
Gesellschaft fur Anlagen- und
Reaktorsicherheit (GRS) mbH
Forschungsgelände
85748 Garching b. München
GERMANY

Mr. **İrfan Koca**
Turkish Atomic Energy Authority
ANAEM, Fen Fakültesi Arkası
Beşevler, Ankara 06100
TURKEY

Prof. **Marzio Marseguerra**
Polytechnic of Milan
Department of Nuclear Engineering
Via Ponzio 34/3
20133 Milano
ITALY

Mr. **Selim Menteşoğlu**
Çekmece Nuclear Research and
Training Center
Havaalanı, P.K. 1
İstanbul
TURKEY

Prof. **Ali Mosleh**
University of Maryland
Materials and Nuclear Engineering
Department
College Park, MD 20742
U.S.A.

Dr. **Ioannis A. Papazoglou**
National Center for Scientific Research
"DEMOKRITOS"
Institute of Nuclear Technology -
Radiation Protection
P. O. Box 60228
GR-15310 Aghia Paraskevi
Athens
GREECE

Dr. **Gareth W. Parry**
Halliburton NUS Corp.
910 Clopper Rd.
Gaithersburg, MD 20877-0962
U.S.A.

Dr. **Ann Patterson-Hine**
NASA Ames Research Center
Mail Stop 269-3
Moffet Field, CA 94035-1000
U.S.A.

Mr. **Miguel Sanchez Perea**
Consejo de Seguridad Nuclear
Justo Dorado 11
Madrid 28020
SPAIN

Mr. **İsmail Hakkı Polat**
1st Air Supply and Maintenance
Center Command
Eskişehir
TURKEY

Mr. **Sam Porter**
Health and Safety Executive
St. Anne's House
Stanley Precinct, Bootle
Merseyside L20 3RA
U.K.

Dr. **Nathan Siu**
Idaho National Engineering Laboratory
Center for Reliability and Risk
Assessment, MS 2405
P. O. Box 1625
Idaho Falls, ID 83415
U.S.A.

Dr. **Carol Smidts**
Commission of the European
Communities Joint Research Center
Institute for Systems Engineering and
Informatics
21020 Ispra (Varese)
ITALY

Dr. **Ayşe Çeçen Tekalp**
The Scientific and Technical Research
Council of Turkey, M. A. M.
Gebze
Kocaeli 41470
TURKEY

Prof. **Bruce K. Walker**
University of Cincinnati
Aerospace Engineering and Engineering
Mechanics Mail Loc. 343
Cincinnati, OH 45221-0343
U.S.A.

Dr. **Peter A. Wieringa**
University of Technology
Department of Mechanical Engineering
Merkelweg 2
2628 CD Delft
THE NETHERLANDS

Mr. John Wreathall
John Wreathall and Co.
4157 Macduff Way
Dublin, OH 43017
U.S.A.

Mr. Mithat Yaman
NETAŞ
Reliability and Component Engineering
Department
Alemdağ Cad. 81244
Ümraniye - Istanbul
TURKEY

Mr. Enrico Zio
Polytechnic of Milan
Department of Nuclear Engineering
Via Ponzio 34/3
20133 Milano
ITALY

NATO ASI Series F

Including Special Programmes on Sensory Systems for Robotic Control (ROB) and on Advanced Educational Technology (AET)

Vol. 47: Advanced Computing Concepts and Techniques in Control Engineering. Edited by M. J. Denham and A. J. Laub. XI, 518 pages. 1988. *(out of print)*

Vol. 48: Mathematical Models for Decision Support. Edited by G. Mitra. IX, 762 pages. 1988.

Vol. 49: Computer Integrated Manufacturing. Edited by I. B. Turksen. VIII, 568 pages. 1988.

Vol. 50: CAD Based Programming for Sensory Robots. Edited by B. Ravani. IX, 565 pages. 1988. *(ROB)*

Vol. 51: Algorithms and Model Formulations in Mathematical Programming. Edited by S. W. Wallace. IX, 190 pages. 1989.

Vol. 52: Sensor Devices and Systems for Robotics. Edited by A. Casals. IX, 362 pages. 1989. *(ROB)*

Vol. 53: Advanced Information Technologies for Industrial Material Flow Systems. Edited by S. Y. Nof and C. L. Moodie. IX, 710 pages. 1989.

Vol. 54: A Reappraisal of the Efficiency of Financial Markets. Edited by R. M. C. Guimarães, B. G. Kingsman and S. J. Taylor. X, 804 pages. 1989.

Vol. 55: Constructive Methods in Computing Science. Edited by M. Broy. VII, 478 pages. 1989.

Vol. 56: Multiple Criteria Decision Making and Risk Analysis Using Microcomputers. Edited by B. Karpak and S. Zionts. VII, 399 pages. 1989.

Vol. 57: Kinematics and Dynamic Issues in Sensor Based Control. Edited by G. E. Taylor. XI, 456 pages. 1990. *(ROB)*

Vol. 58: Highly Redundant Sensing in Robotic Systems. Edited by J. T. Tou and J. G. Balchen. X, 322 pages. 1990. *(ROB)*

Vol. 59: Superconducting Electronics. Edited by H. Weinstock and M. Nisenoff. X, 441 pages. 1989.

Vol. 60: 3D Imaging in Medicine. Algorithms, Systems, Applications. Edited by K. H. Höhne, H. Fuchs and S. M. Pizer. IX, 460 pages. 1990. *(out of print)*

Vol. 61: Knowledge, Data and Computer-Assisted Decisions. Edited by M. Schader and W. Gaul. VIII, 421 pages. 1990.

Vol. 62: Supercomputing. Edited by J. S. Kowalik. X, 425 pages. 1990.

Vol. 63: Traditional and Non-Traditional Robotic Sensors. Edited by T. C. Henderson. VIII, 468 pages. 1990. *(ROB)*

Vol. 64: Sensory Robotics for the Handling of Limp Materials. Edited by P. M. Taylor. IX, 343 pages. 1990. *(ROB)*

Vol. 65: Mapping and Spatial Modelling for Navigation. Edited by L. F. Pau. VIII, 357 pages. 1990. *(ROB)*

Vol. 66: Sensor-Based Robots: Algorithms and Architectures. Edited by C. S. G. Lee. X, 285 pages. 1991. *(ROB)*

Vol. 67: Designing Hypermedia for Learning. Edited by D. H. Jonassen and H. Mandl. XXV, 457 pages. 1990. *(AET)*

Vol. 68: Neurocomputing. Algorithms, Architectures and Applications. Edited by F. Fogelman Soulié and J. Hérault. XI, 455 pages. 1990.

Vol. 69: Real-Time Integration Methods for Mechanical System Simulation. Edited by E. J. Haug and R. C. Deyo. VIII, 352 pages. 1991.

Vol. 70: Numerical Linear Algebra, Digital Signal Processing and Parallel Algorithms. Edited by G. H. Golub and P. Van Dooren. XIII, 729 pages. 1991.

NATO ASI Series F

NATO ASI Series F

NATO ASI Series F